I0067511

Saving American Manufacturing

Saving American Manufacturing

The Fight for Jobs, Opportunity, and National Security

William R. Killingsworth

BEP   BUSINESS EXPERT PRESS

Saving American Manufacturing: The Fight for Jobs, Opportunity, and National Security

Copyright © Business Expert Press, LLC, 2014.

All rights reserved. No part of this publication may be reproduced, stored in a retrieval system, or transmitted in any form or by any means—electronic, mechanical, photocopy, recording, or any other except for brief quotations, not to exceed 400 words, without the prior permission of the publisher.

First published in 2014 by
Business Expert Press, LLC
222 East 46th Street, New York, NY 10017
www.businessexpertpress.com

ISBN-13: 978-1-60649-610-7 (paperback)
ISBN-13: 978-1-60649-611-4 (e-book)

Business Expert Press Economics Collection

Collection ISSN: 2163-761X (print)
Collection ISSN: 2163-7628 (electronic)

Cover and interior design by Exeter Premedia Services Private Ltd, Chennai, India

First edition: 2014

10 9 8 7 6 5 4 3 2 1

Printed in the United States of America.

To my beautiful and amazing wife, Susan, who endured countless hours alone while I did my writing. Her encouragement, patience, and "great dinners" helped to make this possible. Thank you my love!

"It is a capital mistake to theorize before one has data. Insensibly, one begins to twist the facts to suit theories, instead of theories to suit facts."

Sherlock Holmes, A Scandal in Bohemia

Contents

Preface

.... to be independent for the comforts of life we must fabricate them ourselves. We must now place the manufacturer by the side of the agriculturist... He, therefore, who is now against domestic manufacture, must be for reducing us either to dependence on that foreign nation, or to be clothed in skins, and to live like wild beasts in dens and caverns. I am not one of these; experience has taught me that manufacturers are now as necessary to our independence as to our comfort...

—Thomas Jefferson, *Thomas Jefferson Writings*, 1984

Thomas Jefferson astutely recognized the danger of dependence on a foreign nation for essential goods and the importance of manufacturing for the preservation of our independence. For a young country with a nascent manufacturing sector, the dangers to the United States were clear and real. Today our country again faces mounting dependence on a foreign nation and the dangers again appear to be clear and real. The situation is a cause for concern and action.

Perhaps surprisingly, the concern does not regard America's dependency on foreign oil producers. The United States has long worked toward "energy independence" and as a result, with new production techniques and growing efficiencies, monthly domestic production exceeded imports for the first time in two decades in October of 2013 (Bloomberg News 2013). This is, of course, a very positive development for both the economy and our national security. Nevertheless, the United States still imports 35 percent of the petroleum it uses and the trade deficit for oil was a staggering $269 billion for the year 2013 (USA Today 2013). These figures are still a cause for concern and strong motivation for continuing efforts to reduce the oil dependency and the outflow of dollars to other countries.

If not oil, then what dependency is of concern? It's the one to which Thomas Jefferson referred to in the previous quote—manufactured goods. This book examines the concern for manufactured goods that range from computers to communication equipment, critical pharmaceutical components, machine tools, and paper and printing. As Sherlock Holmes advised, consider these facts:

- Between 1999 and 2007, while the U.S. economy was growing, manufacturing lost 3.4 million jobs; another 2 million jobs were lost in the Great Recession between 2007 and 2012 (Bureau of Labor Statistics 2014a).
- Between 1999 and 2007, the number of manufacturing plants and factories dropped by 50,000; another 25,000 establishments were lost during the recession between 2007 and 2012 (Bureau of Labor Statistics 2014a).
- Our trade deficit for manufactured goods was $449 billion in 2013 (ITA Global Patterns of U.S. Trade 2014)! This trade deficit for manufactured goods was $180 billion *in excess* of our $269 billion deficit for oil (ITA Global Patterns of US Trade 2014).
- Our trade deficit for manufactured goods from China alone was $343 billion in 2013, $74 billion greater than our deficit for oil; that's worth repeating—our trade deficit for stuff imported from China is greater than our trade deficit for all the oil we import (ITA Trade with Selected Market 2014)!

Critical industries for national security have been decimated:

- The value of shipments for U.S. manufactured computers and peripheral equipment has fallen from roughly $80 billion in the year 2000 to just $10 billion in 2012 (U.S. Census Bureau 2013) and, over the same time period, the U.S. trade deficit with China for this equipment has soared to $63 billion (ITA Trade with Selected Market 2014).
- The value of shipments for U.S. manufactured communications equipment was $120 billion in 2000 but fell to roughly $40 billion by 2012 (U.S. Census Bureau 2013) and, again over the same time period, the U.S. trade deficit with China for this industry grew to $60 billion (ITA Trade with Selected Market 2014).
- A recent government report found that the United States has lost its ability to manufacture cutting-edge

telecommunications equipment (Office of the Deputy
Assistant Secretary of Defense 2013).

- In a world that is increasingly digitized, networked, and
interconnected, this foreign dependency for computers and
communication equipment is highly troubling and presents
the United States with significant and growing cyber risks.

- In another critical industry, the crucial ingredients for nearly
all antibiotics, steroids, and many other lifesaving drugs are
now made exclusively in China (Harris 2014).

These industries are not alone in succumbing to the manufacturing powerhouse of China. A broad variety of U.S. manufacturing industries have
been impacted and exhibit very similar patterns of stagnation and declines
in employment, plant closings, production declines, capacity stagnation,
and reduced value of shipments.

Warnings of these dangers started before the North American Free
Trade Agreement (NAFTA) with presidential candidate Ross Perot
famously warning of the "giant sucking sound" of jobs going to Mexico.
The warnings grew sharply after NAFTA was enacted and especially after
China was granted Permanent Normal Trade Relations by the U.S. in 2000
and gained membership to the World Trade Organization in 2001. Plant
after plant was closed and jobs were sent overseas. Under heavy lobbying
pressure from multinationals anxious to increase profits by off-shoring jobs
to low-wage countries, the warnings were ignored by U.S. presidents and
Congress. The lobbying message was consistent—"Free trade will create
jobs in the United States." Unfortunately that has *never* been true as foreign
countries often relied on currency manipulation and protectionist barriers to create a non-level playing field. American exports to these countries
have never amounted to much—except, of course, the exports of American
jobs. As a consequence, millions of U.S. manufacturing jobs have been
lost. Wages as a percent of GDP have been falling steadily since the year
2000 and, as might be expected, corporate profits as a percentage of GDP
have been climbing. Unsurprisingly given the trends in wages and profits,
the share of total U.S. income for the top 1 percent has been growing and
the share of income for the other 99 percent has been falling. So-called
"free trade" and the concurrent loss of American manufacturing jobs have
played a substantial role in the growing income inequality in America.

Some might argue that all is well since the unemployment rate has been falling for several years. In May of 2014, the unemployment rate had fallen to 6.3 percent from a high of 10 percent in October, 2009 during the height of the recession (Bureau of Labor Statistics 2014). Does this drop really mean that, in reality, the loss of manufacturing jobs is "much ado about nothing?" No, it does not. A falling unemployment rate only tells one side of the story.

The official unemployment rate includes those unemployed workers who are able to work, have actively looked for work in the past four weeks, but have not found or taken a job or been recalled to a previous job. A person is not counted as unemployed if he or she has gotten frustrated with looking for a job and has given up trying to find work. This implies that the official unemployment rate understates the true rate of unemployment. Moreover, the unemployment rate can also understate the actual unemployment rate because it does not include those who are underemployed, for example, working part-time when they would prefer to be working full-time- or who are working at jobs that are below their skill, educational or wage levels.

Because the unemployment rate does not include many individuals who have given up or are not fully employed, the labor force participation rate is often considered a more indicative measure of employment. This measure is determined as the percentage of the adult population that is working. More precisely, it is the proportion of the civilian noninstitutional (not in college or the Army for example) population aged 16 years and over that is employed. In the year 2000 roughly 64.5 percent of the population in the United States was working. (Bureau of Labor Statistics 2014) That percentage has now fallen to approximately 58.5 percent, a drop of 6 percent. Between the years 1990 and 2000, the number in the civilian noninstitutional population increased by 12.4 percent. The number employed grew even faster, with an increase of 15.2 percent. The number of individuals not in the labor force grew by just 10.3 percent, lagging behind the growth of the population and the number of employed. In other words, jobs were being created faster than the growth of the civilian noninstitutional population. After the year 2000, however, that dynamic changed rather dramatically. The civilian noninstitutional population grew by 16.3 percent from 2000 to 2014, close to the same rate as from 1990 to 2000. The number employed, however, only grew by 6.3 percent Over the same period, the number of individuals not in the labor force, the ones being left behind, grew by 21.3 million workers, an increase of 30.8 percent—a dramatic and heartbreaking increase. The population not in the labor force grew sharply after the year 2000 and now totals 91.5 million citizens of the U.S. This truly impacts the growing income inequality in the United States.

The loss of six million manufacturing jobs absolutely matters. That importance is heightened by the fact that every manufacturing job is estimated to generate from 2.2 to 4.6 additional jobs in support industries. (Wial 2013) Even at the low end, that's an additional 12 million jobs lost for a total loss of 18 million jobs, a significant part of the 21 million that were added to those not in the labor force between 2000 and 2014. It is vital that we save American manufacturing and stop the bleeding of jobs.

Many American manufacturing industries are now struggling to survive. U.S. dependency on foreign goods and trade deficits continue to grow. For the past 20 years the United States has been caught in a vicious cycle, continually weakening our global competitiveness. As American's foreign dependencies for critical products continue to grow, the United States is less able to take forceful actions against countries such as China that manipulate currency and impose non-tariff protectionist barriers. The vicious cycle continues unabated; we are a nation at risk.

Yet even in the face of these remarkably disturbing facts there are still reports and articles that suggest U.S. manufacturing is fine, that productivity is growing rapidly and reducing the need for workers (Reich 2009), and that American manufacturing is poised for a comeback (Morris 2013). The objectives of this book are to paint a detailed picture of the recent history of U.S. manufacturing and to bring awareness to the seriousness of the situation. The book presents data over time on employment, number of establishments, loss of establishments by size, industrial production (IP) and capacity, and balance of payments (BOP) for many specific manufacturing industries. The goal, again as suggested by Sherlock Holmes, is to develop a robust theory based upon the facts, not to first develop a theory and then use selected data for validation and alleged proof. In this case, the facts and the data clearly indicate that U.S. manufacturing industries are in trouble and that foreign dependencies are increasing in vital industries.

A significant comeback is unlikely without strong action given the negative strength of the long-term trends. Entire manufacturing ecosystems and supply chains have left this country. The United States must take a forceful stand to stop and reverse these trends. The danger here is the nature of that stance. The American public is increasingly hostile to free trade. A 2010 poll conducted by the Wall Street Journal and NBC News found "83 percent of blue-collar workers agreed that outsourcing of manufacturing to foreign countries with lower wages was a reason the U.S. economy was struggling and more people weren't being hired; no other factor was so often cited for current economic ills. Among professionals and managers, the sentiment was even stronger: 95 percent of them blamed outsourcing" (Murray 2010). Eighty-four percent of Democrats agreed and 90 percent of Republicans agreed. The danger is that of a populist appeal to unilateral protectionist measures that could ignite a trade war. That is not a promising approach for the United States since much damage could be done to the United States through retaliation.

The United States can begin reversing the downward trends and save American manufacturing through the following five actions and policies:

1. The United States must insist that other countries stop manipulating their currencies and permit the dollar to regain a competitive level. The president and the Executive Branch can do this most readily, but Congress could demonstrate United States resolve through legislation. The United States should first seek voluntary agreement from the currency manipulators to greatly reduce or eliminate their intervention. If the manipulators do not do so, however, the United States should adopt four new policy actions as suggested by Bergsten and Gagnon of the Peterson Institute for International Economics (Bergsten 2012):

   a. Undertake countervailing currency initiatives by buying amounts of their currencies equal to the amounts of dollars they are buying themselves, thus neutralizing the impact on exchange rates;
   b. Tax the earnings on, or restrict further purchases of, dollar assets acquired by intervening to penalize them for building up these positions;

c. Treat manipulated exchange rates as export subsidies for purposes of levying countervailing import duties; and

d. Bring a case against the manipulators in the World Trade Organization that would authorize more wide-ranging trade retaliation.

Bergsten and Gagnon suggest this approach should first be taken against eight of the most significant currency manipulators: China, Denmark, Hong Kong, Korea, Malaysia, Singapore, Switzerland, and Taiwan.

2. The president and Executive Branch must pursue aggressive actions and complaints with the WTO regarding non-tariff protectionist barriers. In 2013, in a very difficult trade environment, the United States filed not a single complaint with the WTO. Given the extent to which the United States is being "beat up" in international trade, this lack of action is difficult to understand. Moreover, multiple complaints from previous years have not been resolved. "While Nero fiddled ...." The United States must aggressively seek resolution of past complaints and continue filing current complaints with this world body.

3. The national government should take a lesson from State governments and actively compete globally for new manufacturing facilities. This competitive stance should include lower income tax rates for manufacturing, tax credits for manufacturing job creation, more rapid depreciation on plant and equipment, permanent R&D tax credits and other incentives. These actions will help level the global playing field that is well tilted against American manufacturing.

4. The U.S. Congress should fund the proposed National Network for Manufacturing Innovation. This network will bridge the so-called "valley of death" between basic research and products made in American factories. Technology, innovation, and creativity have been the bedrock of U.S. economic growth. These institutes have the very real potential for initiating a renaissance in manufacturing and job creation.

5. The Department of Commerce should immediately undertake a competitiveness audit to identify those industries in which America is still globally competitive and near-competitive. This audit will provide valuable insights for investment, R&D, and strategy decisions. Otherwise, actions will be taken in the dark.

If these five actions are forcefully undertaken and if—and it's a big if—Congress and the president can stand up to lobbyists representing multinationals and foreign countries, American manufacturing can be saved, good jobs will be created, income inequality will be reduced, and national security will be strengthened. Polls consistently indicate that the American public and voters believe trade policies have caused massive job losses, yet under the influence of corporate lobbyists, elected officials have not taken corrective actions. It is time for elected officials to stand up for American workers and jobs.

As Thomas Jefferson noted, our independence depends upon it.

Keywords

balance of trade, currency intervention, currency manipulation, imports, income inequality, industrial production, job loss, manufacturing, manufacturing establishments, NAFTA, productivity, unemployment, value of shipments, World Trade Organization

Prologue

As I opened my Christmas gifts, I came to the realization that none of my gifts were made in the United States. A shirt from China, another shirt from Indonesia, a tablet made in China, and a radio-controlled car made in China. Later in the day, when I took some medicine, one pill was made in India and the other was made in Israel. Then when I got on the Internet, I used a router made in China and a laptop made in China. It was so distressing that I had a drink made in Scotland.

I work extensively in manufacturing, helping companies plan, organize, and implement their strategies for supply chain and manufacturing operations. Often I'm called in to help solve supply chain problems and to identify and mitigate supply chain risks. Much of my work has been conducted in the aviation, aerospace, and defense industries. I have also worked with the automotive, pharmaceutical, electronics, and electric power industries. All of these supply chains typically include hundreds of companies and thousands of parts and components. Two years ago, I participated in an effort to analyze critical defense manufacturing supply chains on a Sector-by-Sector, Tier-by-Tier (S2T2) basis (Office of the Deputy Assistant Secretary of Defense 2013). This program was spearheaded by the Manufacturing and Industrial Base Policy office within the Office of the Secretary of Defense. The objectives of this effort were to:

- Map critical supply chains from Original Equipment Manufacturers (OEMs) down through the supply chain tiers all the way to the raw material suppliers.
- Highlight particular parts of the supply chain that depend upon a sole supplier or foreign sources.
- Investigate sources of innovation and design capability for future products.
- Evaluate sources of working capital and investment capital at the various tiers of the supply chain.
- Investigate the level of globalization among suppliers and potential suppliers.

What we found was truly distressing. First, the supply chains were filled with companies that were the sole suppliers of a part or product, and in many cases, there were no alternative suppliers. This is known as a "single point of failure": A part of a system that, if it fails, will stop the entire system from working. This is a high-risk situation for many supply chains in the United States, but it is almost unavoidable because so many U.S. manufacturers have gone out of business. Secondly, we found that innovation is at risk because new concepts and processes are often developed by lower tier suppliers. The R&D activities at these suppliers (often of small size) are now frequently constrained by cash flow limitations. Thirdly, we found that small to medium size manufacturing enterprises often face difficult situations accessing working capital and investment capital. This constrains their ability to acquire new equipment and new technology and maintain global competitiveness. And fourthly, we identified many items and products that could only be acquired from a foreign company—the parts or products were no longer manufactured in the United States.

In my experience, these findings apply to many commercial manufacturing networks in the United States, not just to defense-related products. Because I have seen the diminishing manufacturing base in the United States and have concerns about the implications, I undertook the research presented in this book. A recent article I read was titled "Factory Jobs Are Gone. Get Over It" (Kenny 2014). Well, I can't "get over it." We have serious problems that cannot be ignored. It is my goal to raise awareness of the challenges we now face in the United States and to motivate the adoption of policies that will halt the deterioration of manufacturing in the United States.

Final note: When I went to the office printer to retrieve a printed version of this preface, I had to load paper into the printer. On the wrapper of the ream of paper was a graphic of a kangaroo—looking more closely, I saw that the paper was made in Australia. Importing paper? From Australia? Really? Is it that bad? What can I say? As one of my friends commented: "We've been asleep at the wheel." We must wake up before we're in the ditch.

Bill Killingsworth

Acknowledgments

Many friends and colleagues have been highly supportive in the development of this book and support of my research. First, I want to thank the great people at Business Expert Press: Scott Isenberg, Destiny Hadley, Dr. Phillip Romero, Dr. Jeffrey Edwards, and Sheri Dean. Working with them has been a pleasure. I want to thank MIT Professor David Simchi-Levi and the members of the MIT Forum for Supply Chain Innovation. Their encouragement, guidance, and advice have been invaluable. Prof. Tom Kurfess of Georgia Tech and Prof. Alex Slocum of MIT have also given valuable insights and feedback. Rick Jarman and Rebecca Taylor of the National Center for Manufacturing Sciences contributed on-going guidance regarding manufacturing technology. Special thanks go to John Holzwarth, CFA and partner at OSKR, LLC, for dragging me out of the data morass to higher ground and encouraging me to address the strategic and political issues relating to the decline of manufacturing. My friend Dick Reeves reviewed a draft of the book and raised the conundrum of falling unemployment in the face of manufacturing job losses. John Vickers, Director of the NASA Center for Advanced Manufacturing, has provided many insights into the shrinking industrial base, and Frank De Luca in the Army PEO Missiles and Space shared his valuable strategic perspective. Brett Lambert, Neal Orringer, Adele Ratcliff, and Steve Linder in the Office of the Secretary of Defense have taken an active lead in analyzing the defense industrial base, and I am grateful for their support. Ron Perlman of Holland and Knight generously shared his unique insights into the defense industrial base. Norm Montaño, who fights manufacturing and supply chain challenges daily, has provided important perspectives into today's manufacturing challenges. Ralph Resnick, founding Director of the National Additive Manufacturing Innovation Institute, America Makes, has been very supportive. Harry Moser, founder of the Reshoring Initiative, has shared valuable thoughts and insights. Charles Gilbert at the Federal Reserve provided assistance with data collection and data interpretation. I also want to thank my research sponsors at the Army Aviation and Missile command: Bill Andrews, Artro Whitman, Brian Wood, and Wayne Bruno. I must also thank my friends and colleagues at the University of Alabama in Huntsville: Dr. Steven Messervy, Dr. Richard Rhoades, Dr. Kenneth Sullivan, Chris Sautter, Joe Paxton, Brian Tucker, Teri Martin, and Karen Hancock. Teri Martin deserves special thanks for providing great assistance with data collection and management and for graphical charts. My friends General Jim Rogers, Wally Kirkpatrick, John Vickers, Mike Ward, Peter Beucher, and Ronnie Boles endured a prolonged presentation of my findings and gave valuable feedback and insights. Very special thanks are due to Dr. Wallace Kirkpatrick, Michael Kirkpatrick, and Dave Hemingway of DESE Research for their on-going support and confidence in this research. Dan Tripp and Zac Singh have provided valuable personal friendship and support over many years. Finally, I want to thank my wife Susan for her patience, support, and love during the prolonged birthing of this book.

CHAPTER 1

Background

> Manufacturing is a vital sector of a modern economy and is crucial for national defense and security. Government policies that create a level playing field and enhance global competitiveness of American industry provide the foundation for rebuilding American manufacturing, creating jobs, and securing our nation.

Since 1990, the United States has lost one in every three manufacturing jobs. U.S. employment in manufacturing in 1990 was 17,695,000. By 2011, the number of Americans employed in manufacturing had dropped to 11,734,000—a stunning loss of roughly six million jobs, a 34 percent drop in manufacturing jobs (Bureau of Labor Statistics 2014a). The decline of the U.S. manufacturing industry has been severe and its impact devastating to many American families and communities. Many cities are filled with vacant and abandoned factories. Because of the vast job losses and concerns regarding national security, manufacturing is now a major topic of discussion. Articles appear almost daily in major newspapers and business publications announcing either the growth or demise of American manufacturing: "Moving Back to America" (The Economist 2011); "Is U.S. Manufacturing Making a Comeback" (Sirkin 2011); "Manufacturing Is Surprising Bright Spot in U.S. Economy"

(Norris 2011); "That Giant Sucking Sound of Manufacturing Jobs Going to China" (Worstall 2012); "Worse Than the Great Depression: What Experts Are Missing About American Manufacturing Decline" (Atkinson et al. 2012).

Regardless of the headlines, manufacturing employment has indeed recovered modestly from the depths of the Great Recession (2007–2009). A number of companies have even announced the reshoring of jobs and the building of new plants in the United States. These positive reports include the following examples: NCR is moving production of its ATMs to Columbus, Georgia that will employ 870 people by 2014; Ford is bringing up to 2,000 jobs back to the U.S. in the wake of a favorable agreement with the United Auto Workers (UAW); General Motors (GM) has announced that it will invest two billion dollars to add up to 4,000 jobs at 17 American plants; and Caterpillar has announced a new plant to be built in Athens, Georgia that will employ 1,400 (Morris 2013; The Economist 2011; Sirkin 2011). These developments are certainly positive and will result in several thousand direct manufacturing jobs, and will create many additional manufacturing and service jobs to support these new facilities and their employees. But it must be acknowledged, however, that the total number of jobs within this reshoring group is still a long way from the six million jobs that have been lost. Current debates center around two questions: Should we care? And, if so, what should we do?

At least four arguments are being made that the demise of U.S. manufacturing either does not matter or should not be a focus of government policy. The first argument is that the job losses are the result of higher productivity in manufacturing. Robert Reich, former secretary of labor, has argued that manufacturing job loss should not be a concern because it results from rapid productivity growth and that should be good for our global competitiveness (Reich 2009). The displaced workers, over time, will find jobs in the service sector. This shift is seen as a natural development for an evolving, modern economy. Numerous studies, however, have shown the fallacy of this argument and demonstrated that the historical U.S. job losses are not the result of massive gains in productivity (Atkinson 2012; Mandel 2012). The second argument suggests that U.S. manufacturing is in trouble because U.S.

wages are too high to be internationally competitive. Again, numerous studies have shown robust manufacturing industries exist in countries such as Germany which have higher wages than the United States (Atkinson 2012). Indeed, the massive manufacturing job losses in the United States do not have a parallel in these countries. The third argument is that manufacturing does not matter and that the United States can become a design, knowledge, and intellectual property economy with manufacturing done elsewhere. Again, research has shown that design and innovation follow manufacturing. That is, if manufacturing is moved offshore, designers tend to lose touch with processes and innovations, and design will then be done closer to manufacturing, and design will also be lost (MIT 2013). The fourth argument is that the government should not be picking winners and losers. In other words, why should the government favor manufacturing industries over, say, the fast food industry? Obesity concerns aside, there are compelling reasons why manufacturing matters.

In a Brookings Institution report, Helper, Krueger, and Wial (2012) lay out cogent and compelling reasons why manufacturing matters:

- Manufacturing provides high-wage jobs, especially for workers who would otherwise earn the lowest wages;
- Manufacturing is the major source of commercial innovation and is essential for innovation in the service sector;
- Manufacturing can make a major contribution to the nation's trade deficit; and
- Manufacturing makes a disproportionately large contribution to environmental sustainability.

It might be noted that this list omits contributions to national defense and security. Others, however, do include this important aspect of manufacturing.

In the report "The Case for a National Manufacturing Strategy," the Information Technology and Innovation Foundation lists five key reasons why manufacturing plays a critical role in the U.S. economy (Ezell and Atkinson 2011):

1. It will be extremely difficult for the United States to balance its trade account without a healthy manufacturing sector.
2. Manufacturing is a key driver of overall job growth and an important source of middle-class jobs for individuals at many skill levels.
3. Manufacturing is vital to U.S. national security.
4. Manufacturing is the principal source of research and development (R&D) and innovation activity.
5. The manufacturing and service sectors are inseparable and complementary.

Manufacturing makes many economic contributions and provides essential products for our nation's security, defense, and critical economic and infrastructure functioning. The defense of the United States requires a broad variety of manufactured products: Aircraft, motor vehicles, computers, helicopters, rocket motors, communication equipment, satellites, and the list goes on and on. All of these items require a broad and deep manufacturing supply chain for their production. Foundation manufacturing industries for the final products include primary metals, foundries and forgings, chemicals, electronic components, pharmaceuticals, glass and ceramics, industrial machinery, and so forth. It is naïve to believe that we can always plan to have friendly sources for all elements in the defense and security related supply chain. For multiple reasons, manufacturing matters and it is important that U.S. manufacturing be strengthened and rebuilt.

This process requires a diagnosis and understanding of the predicament—What has happened over the past 20 years and how has U.S. manufacturing arrived at this point? Have all manufacturing industries been impacted equally? Are there patterns in the declines? These and other questions are addressed in the following chapters.

A note is appropriate here regarding the data and analysis to be presented. Economic, industrial, and international trade systems evolve, grow, and decay over years, decades even. Manufacturing industries rise and fall over time; trade patterns display dynamics that occur over years. Short-term comparisons, for example year to year, do not reveal the long-term trends and patterns. The analysis in this book does not focus on month to month changes, quarter to quarter changes, or even year to year changes—it focuses on, and presents, the dynamics that evolve over two to three decades. This enables the long-term system dynamics of the industrial, economic, and trade systems to be revealed. And indeed, the data presented in the following chapters reveal very troubling long-term trends regarding U.S. manufacturing. A sobering note is that these long-term patterns are not easily redirected in the short-term; it's akin to turning a giant container ship. It requires considerable time to change course. A comprehensive strategy incorporating innovative technologies, trade policies, workforce development, and tax reform must be developed and implemented in the near-term to save American manufacturing. A number of critical U.S. manufacturing industries appear to be on the road to extinction.

CHAPTER 2

Historical Job Losses in Manufacturing

Introduction

Figures 2.1a and 2.1b present the total number of employees in U.S. manufacturing from 1990 to 2012 (employment data are from the Bureau of Labor Statistics (BLS) 2014). Employment is defined by BLS as the total number of persons on establishment payrolls employed full- or part-time who received pay for any part of the pay period that includes the 12th day of the month. Temporary and intermittent employees are included. In Figures 2.1a and 2.1b, the shaded areas represent the three recessions that occurred over this period (NBER 2010).

> July 1990—March 1991
> March 2001—November 2001
> December 2007—June 2009

As may be seen, job losses are distinctly associated with the recessions. Following the 1990 recession, however, the number of jobs in manufacturing almost recovered to the 1990 level by 1998. The recession of 2001, however, led to a substantial drop and those losses were not recovered. Another large drop occurred with the 2007–2009 recession, and a modest uptick in manufacturing employment is now evident. Figure 2.1b highlights a pattern that will be discussed in this chapter in which employment in industry after industry begins a precipitous decline around the year 2000.

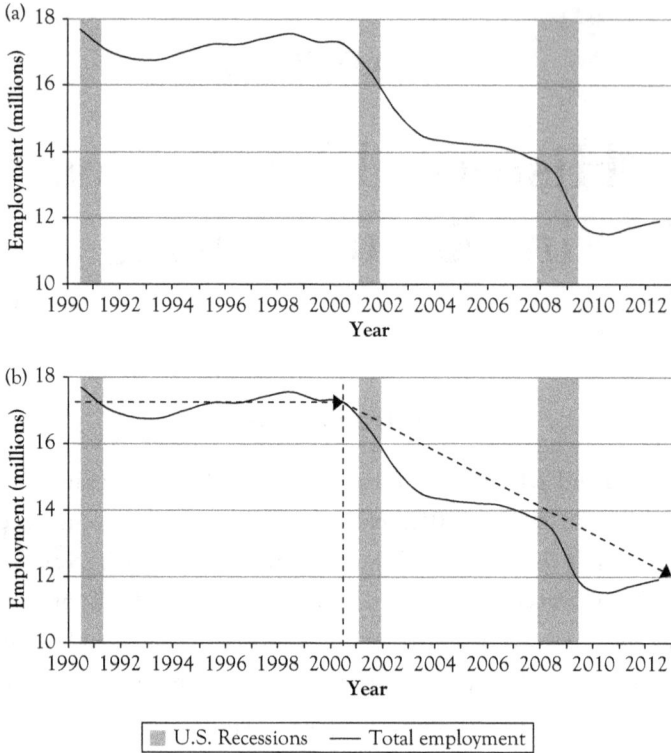

Figure 2.1 Total employment for U.S. manufacturing sector
Source: Bureau of Labor Statistics (2014a).

As a first step, we must address the fundamental question "What do we mean by manufacturing?" The term includes everything ranging from the manufacturing of Doritos to Chevy Malibus to Viagra to plywood to clothing. Strategy and policy development requires segmentation of the all-encompassing term manufacturing. The common basic division is between manufacturing of nondurable goods (goods that are not expected to last for three years, for example, food, beverages, textiles, apparel, chemicals, and pharmaceuticals) and the manufacturing of durable goods (goods that are expected to last for 3 or more years, for example, motor vehicles, aircraft, castings, forgings, computers, semiconductors, and agricultural and mining equipment). Figure 2.2 presents the total employment for the manufacturing of durable goods and for nondurable goods between 1990 and 2012. As may be seen, the two historical employment

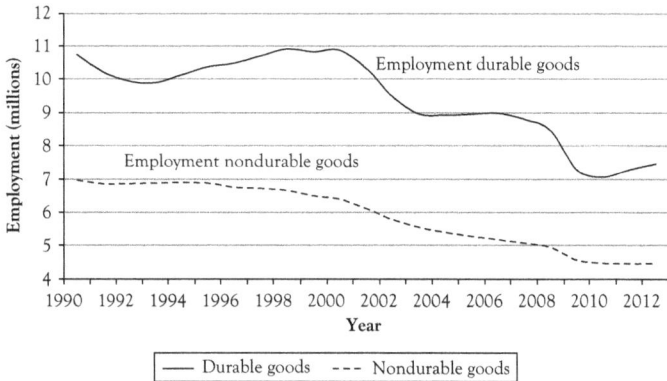

Figure 2.2 Total employment for manufacturing of durable goods and nondurable goods
Source: Bureau of Labor Statistics (2014a).

trend lines look quite different: The nondurable employment begins a steady decline in 1996 and the decline accelerates somewhat following the recession of 2001, and then the nondurable manufacturing employment experiences another drop during the recession of 2007–2009 and then appears to continue on a downward trend. On the other hand, the employment for the manufacture of durable goods shows a decline associated with the 1990 recession but then a full recovery in the number of employees by 1997. The recession of 2001, however, led to a sharp decline with no recovery of lost jobs. Somewhat similarly, the recession of 2007–2009 again led to a major reduction of employees in the manufacturing of durable goods, but a slight uptick in employment is visible in 2011. With this rebound, the manufacturing of durable goods appears to be a potential area for manufacturing recovery and growth, but the question is—where are the important and favorable opportunities for possible growth strategies and policies? This question requires an analysis of specific manufacturing subsectors and industries.

Manufacturing Employment for Nondurable Goods

Figures 2.3a, 2.3b, and 2.3c present the employment by subsector in the manufacturing of nondurable goods. Although every subsector has declined over the period, the most dramatic decline is in apparel.

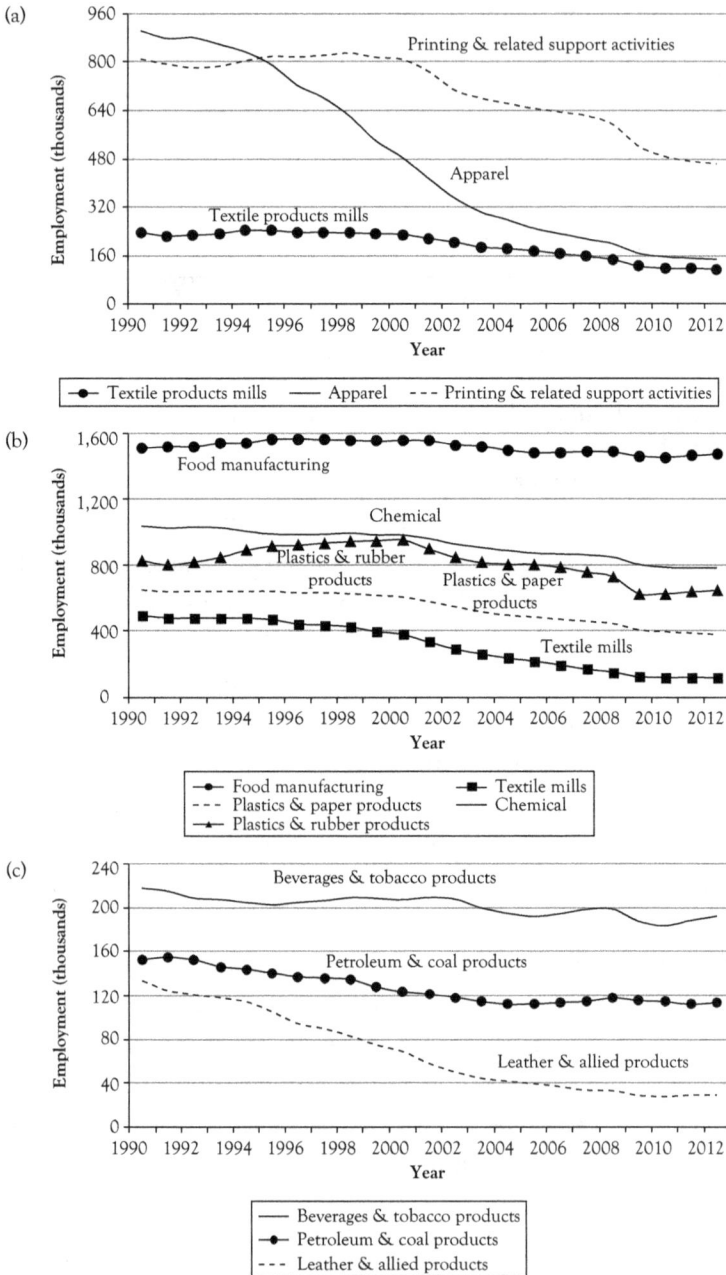

(a)

(b)

(c)

Figure 2.3 Nondurable goods employment by subsector

Source: Bureau of Labor Statistics (2014a).

Table 2.1 presents in tabular form the employment data contained within Figures 2.3a, 2.3b, and 2.3c for nondurable manufacturing subsectors, but additionally provides employment for the subsectors, the loss of employment, and the percentage loss of jobs between 1990 and 2012.

As seen in Table 2.1, employment in the apparel subsector dropped from 902,800 in 1990 to 148,100 in 2012, a drop of 754,700, representing an 83.6 percent loss of employment. The second largest drop in nondurable manufacturing employment was in the textile mills subsector. This subsector declined from an employment of 491,800 in 1990 to 118,000 in 2012, a decline of 373,800 and a loss of 76.0 percent. The third largest drop in nondurable manufacturing employment was in the textile product mills subsector. This subsector declined from an employment of 235,600 in 1990 to 116,600 in 2012, a decline of 119,000 and a loss of 50.5 percent. These three subsectors related to apparel and textile mills altogether accounted for a total employment loss of 1,247,500. Figures 2.4a, 2.4b, and 2.4c graphically present the dramatic loss of employment in these industries. In terms of strategy and potential government policy, the jobs in apparel and textiles are almost certainly gone for good with little hope for reshoring or rebuilding (see Article 2.1). It is very hard to revive an industry after its sales, employment, and facilities have declined so severely. The apparel and textile industries have a history of searching for low wages moving from New England to the Southeast United States and most recently to countries such as China and Indonesia, and now even leaving China for countries such as Vietnam. Government policies and incentives are unlikely to create U.S. jobs in these subsectors.

Within the nondurable goods sector, the chemicals subsector is important and interesting. Figures 2.5a and 2.5b present the employment history for the industries within the chemicals subsector. As may be seen in Table 2.1 and Figures 2.5a and 2.5b, employment for the chemicals subsector dropped from 1,035,700 in 1990 to 783,600 in 2012, a loss of 24.3 percent. However, as may be seen readily in Figures 2.5a and 2.5b, this overall decline for the chemicals subsector is the result of a substantial increase in one industry and declines in the other industries. Specifically, employment in the pharmaceuticals and medicines industry increased from 207,200 to 270,800, a gain of 30.7 percent. On the other hand, the

Table 2.1 Manufacturing of nondurable goods (Employment and losses) (Thousands)

NAICS	Sector Name	Employment		Employment	
		1990	2012	Loss	% Loss
	Nondurable Goods Total	6,958.0	4,456.0	−2,502.0	−35.96
311	**Food manufacturing**	**1507.3**	**1468.7**	**−38.6**	**−2.56**
3111	Animal food	57.0	53.3	−3.7	−6.49
3112	Grain and oilseed milling	71.3	60.2	−11.1	−15.57
3113	Sugar and confectionery products	99.4	66.8	−32.6	−32.80
3114	Fruit and vegetable preserving and specialty	218.1	169.6	−48.5	−22.24
3115	Dairy products	144.5	135.8	−8.7	−6.02
3116	Animal slaughtering and processing	427.4	484.8	57.4	13.43
3117	Seafood product preparation and packaging	54.2	39.4	−14.8	−27.31
3118	Bakeries and tortilla manufacturing	292.1	284.5	−7.6	−2.60
3119	Other food products	143.2	174.3	31.1	21.72
312	**Beverages and tobacco products**	**217.7**	**192.2**	**−25.5**	**−11.71**
3121	Beverages	172.7	177.8	5.1	2.95
3122	Tobacco and tobacco products	45.0	14.4	−30.6	−68.00
313	**Textile mills**	**491.8**	**118.0**	**−373.8**	**−76.01**
3131	Fiber, yarn, and thread mills	101.7	28.0	−73.7	−72.47
3132	Fabric mills	270.2	55.2	−215.0	−79.57
3133	Textile and fabric finishing mills	119.9	34.8	−85.1	−70.98
314	**Textile product mills**	**235.6**	**116.6**	**−119.0**	**−50.51**
3141	Textile furnishings mills	126.9	52.1	−74.8	−58.94
3149	Other textile product mills	108.7	64.4	−44.3	−40.75
315	**Apparel**	**902.8**	**148.1**	**−754.7**	**−83.60**
3152	Cut and sew apparel	749.6	122.1	−627.5	−83.71
3151 & 3159	All other apparel manufacturing	153.2	26.1	−127.1	−82.96
316	**Leather and allied products**	**133.2**	**29.4**	**−103.8**	**−77.93**
322	**Paper and paper products**	**647.2**	**379.0**	**−268.2**	**−41.44**
3221	Pulp, paper, and paperboard mills	238.3	108.2	−130.1	−54.60

Table 2.1 (Continued)

NAICS	Sector Name	Employment 1990	Employment 2012	Employment Loss	Employment % Loss
3222	Converted paper products	408.9	270.8	−138.1	−33.77
323	**Printing and related support activities**	**808.5**	**462.1**	**−346.4**	**−42.84**
324	**Petroleum and coal products**	**152.8**	**113.2**	**−39.6**	**−25.92**
325	**Chemicals**	**1,035.7**	**783.6**	**−252.1**	**−24.34**
3251	Basic chemicals	249.1	142.9	−106.2	−42.63
3252	Resin, rubber, and artificial fibers	158.0	92.3	−65.7	−41.58
3253	Agricultural chemicals	52.4	36.8	−15.6	−29.77
3254	Pharmaceuticals and medicines	207.2	270.8	63.6	30.69
3255	Paints, coatings, and adhesives	84.5	56.7	−27.8	−32.90
3256	Soaps, cleaning compounds, and toiletries	131.6	103.4	−28.2	−21.43
3259	Other chemical products and preparations	153.0	80.8	−72.2	−47.19
326	**Plastics and rubber products**	**824.8**	**645.2**	**−179.6**	**−21.77**
3261	Plastics products	617.9	515.9	−102.0	−16.51
3262	Rubber products	206.9	129.2	−77.7	−37.55

Source: Bureau of Labor Statistics (2014a).

important industries of basic chemicals, resin, rubber and artificial fibers, and agricultural chemicals showed declines of 42.6 percent, 41.6 percent, and 29.8 percent. These three industries are important manufacturing capabilities for the national well-being and security. Recent events (see Article 2.2) have highlighted the dependence of the United States on a variety of imported basic chemicals.

Employment in the Manufacturing of Durable Goods

Figures 2.6a through 2.6d present employment by subsectors in the manufacture of durable goods. As may be seen in Figure 2.6a, the four largest subsectors in terms of employment, transportation equipment, computer and electronic products, fabricated metal products, and machinery, all

(a)

(b)

(c)

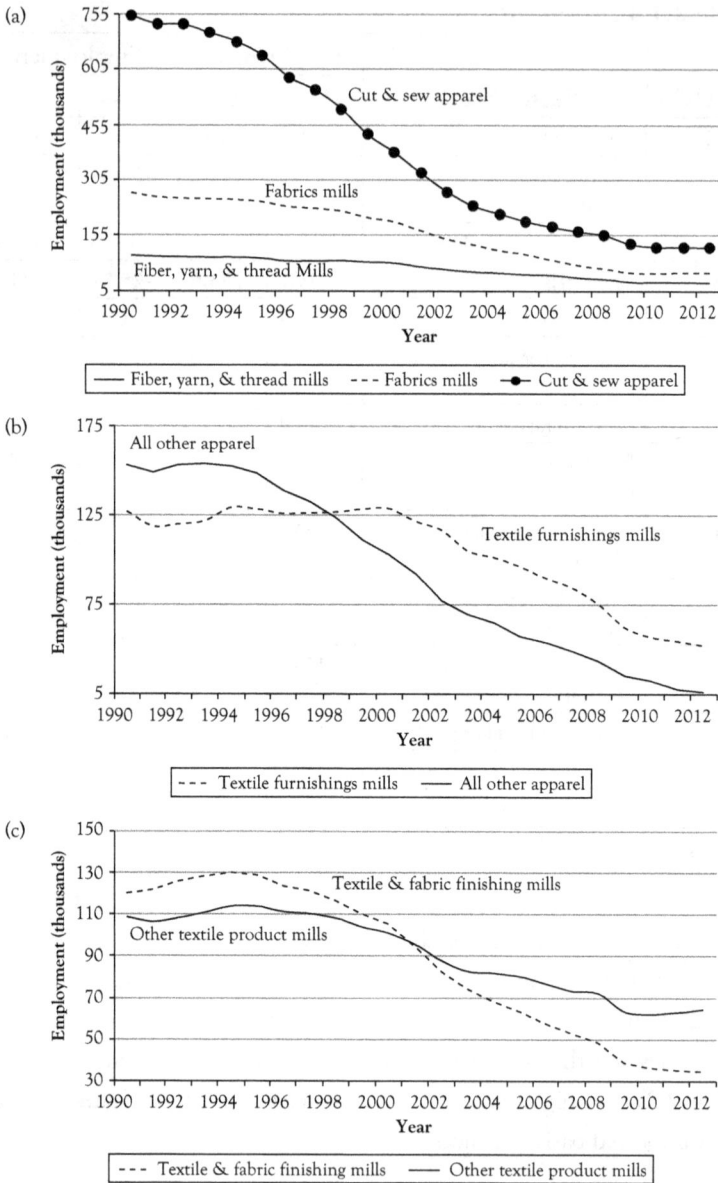

Figure 2.4 Nondurable goods—employment in apparel and textile industries

Source: Bureau of Labor Statistics (2014a).

Article 2.1. U.S. Clothing and Textile Trade with China and the World: Trends Since the End of Quotas; Congressional Research Service, July 10, 2007

At the start of the years 1995, 1998, 2002 and 2005, parties to the Agreement on Textiles and Clothing (ATC) would eliminate quotas for a prescribed percentage of their volume of trade in clothing and textiles. In addition, for those products still subject to quotas, parties to the ATC would increase the quotas by a prescribed percentage, thereby opening their domestic markets to more imported goods. The ATC also required that products from different categories—textiles and clothing, wool, cotton or man-made fibers, etc.—be included in each of the four stages of the quota phase-out, in part to make it more difficult to protect a particular segment of the clothing and textile industry during the transition. While the quota phase-out process appeared relatively gradual in theory, it was relatively abrupt in practice. By selecting less traded products and products with under-utilized quotas for integration in the first three stages, market watchers maintain the United States and other nations were able to prolong the period of protection for product categories where domestic manufacturers held a larger market share until the final stage. Industry analysts, at times, referred to the final quota phase-out on January 1, 2005 as a "cliff," when the quota on the most of the more frequently traded products and the products where existing quotas were typically fully utilized would be lifted.

have very similar historical patterns of job loss—a decline associated with the recession of 1990–91, followed by a strong recovery of employment, then a sharp decline in employment associated with the recession of 2001, followed by either modest recovery or decline, then another sharp reduction associated with the recession of 2007–09. Each of these four subsectors then showed modest increases in the number of employees in 2012. The remaining subsectors also experienced similar trend patterns. Table 2.2 presents, for the durable goods sector and subsectors, the number of employees in 1990 and 2012, the loss of employees, and the percentage decline.

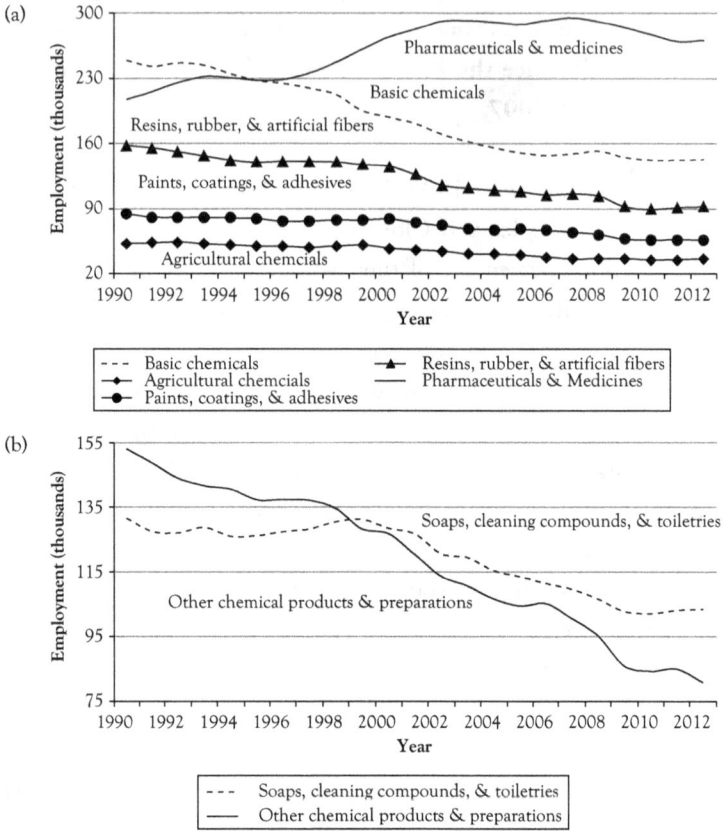

Figure 2.5 *Nondurable goods—employment in chemicals industries*
Source: Bureau of Labor Statistics (2014a).

**Article 2.2. Suppliers race against clock to avoid resin shortage;
Automotive News, April 17, 2012**

Hundreds of representatives from the auto industry are gathering
today in suburban Detroit to secure alternative sources of a key resin
and head off a potentially devastating shortage.

The specter of the resin shortage comes as the auto industry climbs
back after the economic downturn and as some companies are emerg-
ing from the effects of last year's earthquake and tsunami in Japan
and flooding in Thailand. Supplies of the resin—known as PA-12
and used to make fuel tanks, brake components and seat fabrics—are

tightening after a March 31 explosion at an Evonik Industries AG plant in Germany that killed two employees.

Evonik and French competitor Arkema SA account for about half of the world's supply of PA-12, according to research firm IHS Automotive. "It is now clear that a significant portion of the global production capacity of PA-12 (nylon 12) has been compromised," the Automotive Industry Action Group, an industry trade association, said in a statement.

The supply of PA-12 was tightening even before the Evonik accident because the resin also is used in solar panel production, IHS said. Robust demand for the resin has pushed prices higher, and many suppliers and automakers have been searching for alternative materials, according to IHS.

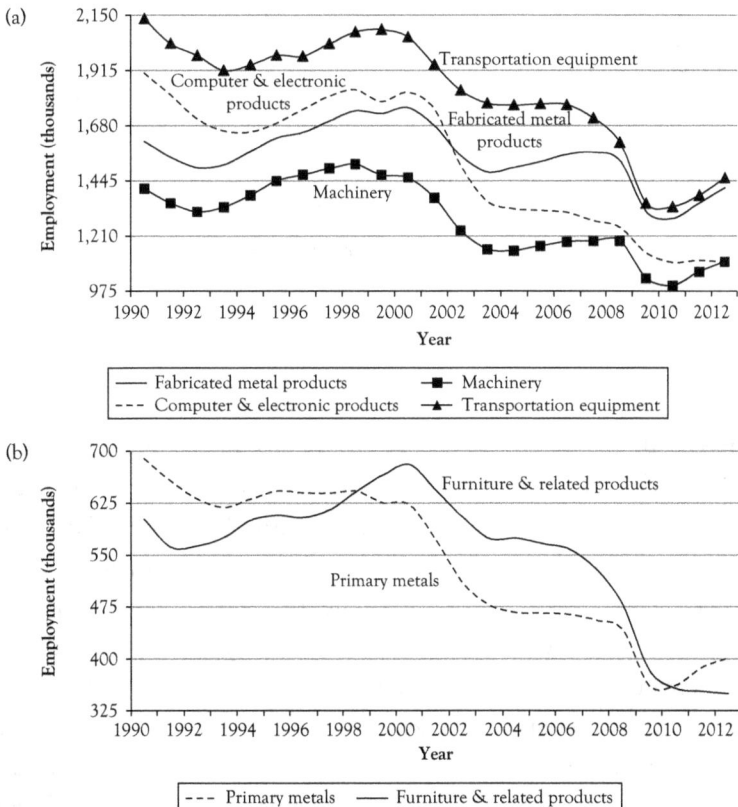

Figure 2.6 (Continued)

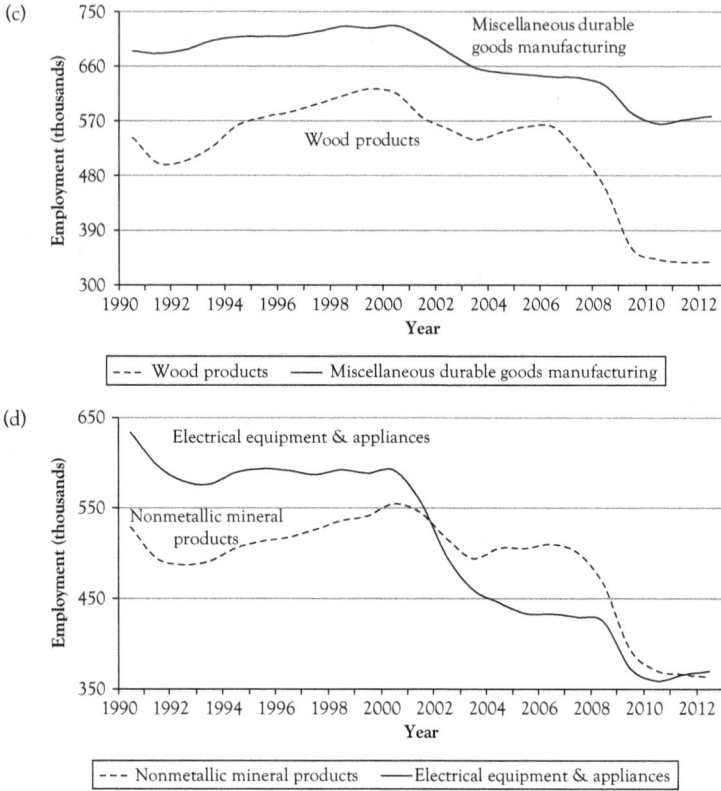

Figure 2.6 *Durable goods manufacturing employment by subsectors*
Source: Bureau of Labor Statistics (2014a).

It is important to note that national well-being and security are closely aligned with six of the durable goods subsectors. It is hard to imagine our ability to defend our country or to have a vital economy without each of these subsectors in a robust condition. These six subsectors are: transportation equipment, fabricated metal products, computers and electronic products, machinery, primary metals, and electrical equipment and appliances. Strong arguments can easily be made for why it is important that the United States have robust manufacturing capabilities in these areas. These subsectors, however, shed 2,547,400 employees between 1990 and 2012.

Figures 2.7a and 2.7b through Figure 2.12 present the historical employment trend lines for the industries in these six subsectors. Figures 2.7a and 2.7b present the number of employees for each industry

Table 2.2 Manufacturing of durable goods (Employment and losses)
(Thousands)

NAICS	Sector Name	Employment		Employment	
		1990	2012	Loss	% Loss
	Durable Goods Total	10,737.0	7,462.0	−3,275.0	−30.50
321	**Wood products**	**543.0**	**337.9**	**−205.1**	**−37.77**
3211	Sawmills & wood preservation	148.1	84.3	−63.8	−43.08
3212	Plywood & engineered wood products	95.5	63.8	−31.7	−33.19
3219	Other wood products	299.4	189.8	−109.6	−36.61
327	**Nonmetallic mineral products**	**528.4**	**363.8**	**−164.6**	**−31.15**
3271	Clay products & refractories	83.6	40.6	−43.0	−51.44
3272	Glass & glass products	152.3	80.0	−72.3	−47.47
3273	Cement & concrete products	194.9	161.6	−33.3	−17.09
3274 & 3279	Lime, gypsum, & other nonmetallic mineral products	97.6	81.6	−16.0	−16.39
331	**Primary metals**	**688.6**	**401.8**	**−286.8**	**−41.65**
3311	Iron & steel mills & ferroalloy production	186.8	93.6	−93.2	−49.89
3312	Steel products from purchased steel	70.4	60.2	−10.2	−14.49
3313	Alumina & aluminum production	108.4	59.5	−48.9	−45.11
3314	Other nonferrous metal production	109.1	61.6	−47.5	−43.54
3315	Foundries	213.9	126.9	−87.0	−40.67
332	**Fabricated metal products**	**1,610.0**	**1,411.3**	**−198.7**	**−12.34**
3321	Forging & stamping	128.1	99.0	−29.1	−22.72
3322	Cutlery & hand tools	78.8	39.6	−39.2	−49.75
3323	Architectural & structural metals	356.8	341.4	−15.4	−4.32
3324	Boilers, tanks, & shipping containers	117.3	96.5	−20.8	−17.73
3325	Hardware	57.2	25.0	−32.2	−56.29
3326	Spring & wire products	77.5	41.6	−35.9	−46.32
3327	Machine shops & threaded products	308.5	362.4	53.9	17.47
3328	Coating, engraving, & heat treating metals	142.5	136.1	−6.4	−4.49

(Continued)

Table 2.2 (Continued)

NAICS	Sector Name	Employment 1990	2012	Employment Loss	% Loss
3329	Other fabricated metal products	343.5	269.9	−73.6	−21.43
333	**Machinery**	**1,409.8**	**1,098.2**	**−311.6**	**−22.10**
3331	Agricultural, construction, & mining machinery	228.7	245.0	16.3	7.13
3332	Industrial machinery	152.5	105.3	−47.2	−30.95
3333	Commercial & service industry machinery	146.7	88.5	−58.2	−39.67
3334	HVAC & commercial refrigeration equipment	165.1	127.7	−37.4	−22.65
3335	Metalworking machinery	266.7	177.1	−89.6	−33.60
3336	Turbine & power transmission equipment	114.1	102.4	−11.7	−10.25
3339	Other general purpose machinery	336.2	252.2	−84.0	−24.99
334	**Computer and electronic products**	**1,902.5**	**1,093.7**	**−808.8**	**−42.51**
3341	Computer & peripheral equipment	367.4	158.6	−208.8	−56.83
3342	Communications equipment	223.0	109.5	−113.5	−50.90
3343	Audio & video equipment	60.1	19.9	−40.2	−66.89
3344	Semiconductors & electronic components	574.0	384.4	−189.6	−33.03
3345	Electronic instruments	634.8	400.4	−234.4	−36.93
3346	Magnetic media manufacturing & reproduction	43.3	21.0	−22.3	−51.50
335	**Electrical equipment and appliances**	**633.1**	**370.1**	**−263.0**	**−41.54**
3351	Electric lighting equipment	80.8	45.8	−35.0	−43.32
3352	Household appliances	113.7	54.6	−59.1	−51.98
3353	Electrical equipment	243.6	142.9	−100.7	−41.34
3359	Other electrical equipment & components	195.0	126.9	−68.1	−34.92
336	**Transportation equipment**	**2,134.5**	**1,456.0**	**−678.5**	**−31.79**
3361	Motor vehicles	271.4	168.0	−103.4	−38.10
3362	Motor vehicle bodies & trailers	129.8	125.4	−4.4	−3.39

Table 2.2 (Continued)

NAICS	Sector Name	Employment		Employment	
		1990	2012	Loss	% Loss
3363	Motor vehicle parts	653.0	479.6	−173.4	−26.55
3364	Aerospace products & parts	840.7	497.4	−343.3	−40.84
3366	Ship & boat building	173.7	129.3	−44.4	−25.56
3365 & 3369	Railroad rolling stock & other transportation equipment	65.9	56.4	−9.5	−14.42
337	**Furniture and related products**	**601.7**	**350.1**	**−251.6**	**−41.81**
3371	Household & institutional furniture	398.2	216.5	−181.7	−45.63
3372	Office furniture & fixtures	156.3	98.9	−57.4	−36.72
3379	Other furniture-related products	47.2	34.6	−12.6	−26.69
339	**Miscellaneous manufacturing**	**685.7**	**579.5**	**−106.2**	**−15.49**
3391	Medical equipment and supplies	283.2	311.2	28.0	9.89
3399	Other miscellaneous manufacturing	402.5	268.3	−134.2	−33.34

Source: Bureau of Labor Statistics (2014a).

in the transportation equipment subsector. After the year 2000, there has been a precipitous decline in employment for manufacturing motor vehicle parts.

The most dramatic loss of employees, however, in Figure 2.7a is in the aerospace products and parts industry, dropping from 840,700 in 1990 to 497,400 in 2012. Most of this decline occurred during the early nineties which was a period of massive reductions in defense spending as well as major consolidation in the industry. In 1998, the U.S. General Accounting Office (GAO) Report 98-141 noted that between 1990 and 1998, the number of manufacturers of Department of Defense fixed wing aircraft dropped from 8 to 3, the number of manufacturers of expendable launch vehicles dropped from 6 to 2, and the number of manufacturers of tactical missiles dropped from 13 to 4. Table 2.3 presents the consolidation from the GAO report, but updates the report as of 2012. As may be seen, additional consolidation has occurred since 1998 with

(a)

Employment (thousands)

875

726

577

428

279

130

1990 1992 1994 1996 1998 2000 2002 2004 2006 2008 2010 2012
Year

Motor vehicle parts

Aerospace products & parts

Motor vehicles

●—● Motor vehicles   —— Motor vehicle parts   - - - Aerospace products & parts

(b)

Employment (thousands)

190

150

110

70

30

1990 1992 1994 1996 1998 2000 2002 2004 2006 2008 2010 2012
Year

Motor vehicle
bodies & trailers

Ship & boat building

Railroad rolling stock & other transportation

—— Motor vehicle bodies & trailers
- - - Ship & boat building
●—● Railroad rolling stock & other transportation

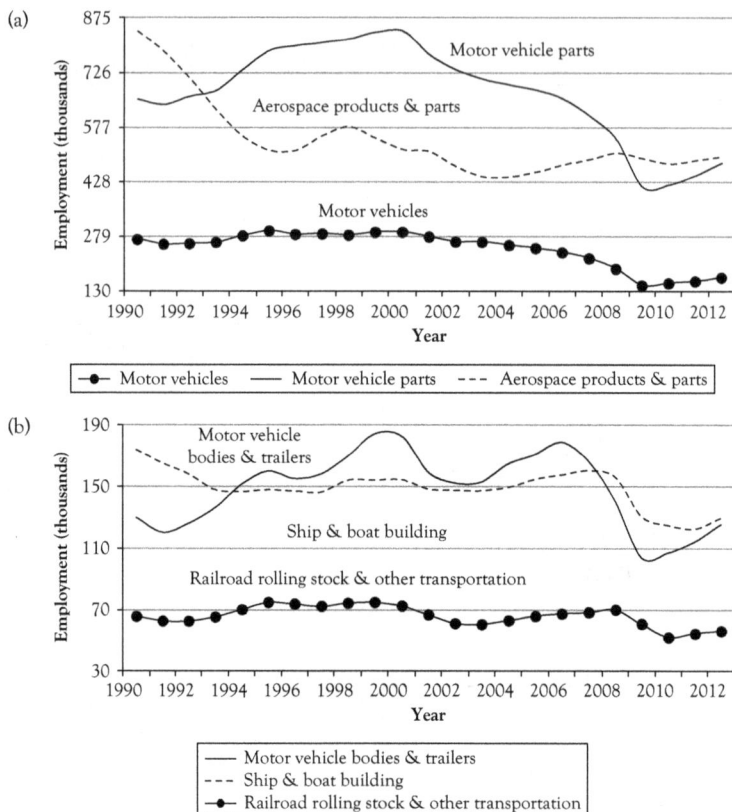

Figure 2.7 Durable goods—employment in transportation equipment industries

companies being purchased or simply going out of business. This much smaller industry is now very concerned about the impacts of looming budget cuts on the remaining companies and employees. Rocket propulsion is an example. In the early 2000s, there were eight companies in the propulsion business: Aerojet, Thiokol, Hercules, Pratt & Whitney Chemical Systems (San Jose, CA), Pratt & Whitney Space Propulsion (West Palm Beach, FL), Rocketdyne, Atlantic Research Corporation (ARC), and Rocket Research. Now there are only two: Aerojet and ATK. Rocketdyne and Rocketdyne's parent, United Technologies Corporation announced (Pasztor 2011) interest in selling Rocketdyne, citing the increased fragility of the U.S. space industrial base, suggesting even more contraction is on the horizon. Rocketdyne was subsequently acquired by Aerojet. Recent articles (see Article 2.3) indicate on-going job losses and likely plant closings in the aerospace industry.

Figures 2.8a and 2.8b present the employment for the industries in the computer and electronic products subsector. As may be seen, job losses have been substantial (see Article 2.4). The manufacturing of semi-conductors and electronic components lost 189,600 jobs between 1990 and 2012, a loss of 33 percent. Employment of electronic instruments has dropped by 234,400, a 37 percent drop. The employment in manufacturing of computers and peripheral equipment has dropped by 208,800, a loss of 57 percent, the manufacturing of communications equipment has dropped by 113,500, a drop of 51 percent. In this digital age, the loss

Table 2.3. GAO Report 98–141, April 1998; Defense Industry Consolidation: 1990–1998

Prime Contractors in Defense Market Sectors			
Sector	Reduction in contractors	1990 contractors	1998 contractors
Tactical missiles	13 to 4	Boeing Ford Aerospace General Dynamics Hughes Lockheed Loral LTV Martin Marietta McDonnell Douglas Northrop Raytheon Rockwell Texas Instruments	Boeing Lockheed Martin Northrop Grumman Raytheon
Fixed-wing aircraft	8 to 3	Boeing General Dynamics Grumman Lockheed LTV-Aircraft McDonnell Douglas Northrop Rockwell	Boeing Lockheed Martin Northrop Grumman
Expandable launch vehicles	6 to 2	Boeing General Dynamics Lockheed Martin Marietta McDonnell Douglas Rockwell	Boeing Lockheed Martin

(Continued)

Table 2.3. (Continued)

Prime Contractors in Defense Market Sectors			
Sector	Reduction in contractors	1990 contractors	1998 contractors
Satellites	8 to 5 1990–2012: 8 to 4	Boeing General Electric Hughes Lockheed Loral Martin Marietta TRW Rockwell	Boeing Lockheed Martin Hughes (Boeing) Loral Space Systems TRW (Northrop)

Surface ships	8 to 5
	1990–2012:
	8 to 2
		Avondale Industries
		Bath Iron Works
		Bethlehem Steel
		Ingalls Shipbuilding
		NASSCO
		Newport News
		  Shipbuilding
		Tacoma
		Tampa
			Avondale Industries
			General Dynamics
			  (Bath Iron Works)
			Ingalls Shipbuilding
			(Huntington Ingalls)
			NASSCO (General
			Dynamics
			Newport News
			(Huntington Ingalls)
			  Shipbuilding

Tactical wheeled vehicles	6 to 4 1990–2012: 6 to 3	AM General Harsco (BMY) GM Canada Oskosh Stewart & Stevenson Teledyne Cont. Motors	AM General GM Canada (General Dynamics) Oskosh Stewart & Stevenson
Tracked combat vehicles	3 to 2	FMC General Dynamics Harsco (BMY)	General Dynamics United Defense LP
Strategic missiles	3 to 2	Boeing Lockheed Martin Marietta	Boeing Lockheed Martin
Torpedoes	3 to 2	Alliant Tech Systems Hughes Westinghouse	Northrop Grumman Raytheon
Rotary wing aircraft	4 to 3	Bell Helicopters Boeing McDonnell Douglas Sikorsky	Bell Helicopters Boeing Sikorsky

> **Article 2.3. Boeing cutting 900 jobs at Long Beach C-17 plant; The sprawling factory, barring an unlikely rise in demand, is expected to shut down completely by the end of next year; W.J. Hennigan, Los Angeles Times, January 20, 2011**
>
> Time is running out at Southern California's last major conventional aircraft factory.
>
> Citing declining orders for its C-17 cargo planes, Boeing Co. said it was cutting 900 of the 3,700 jobs at its sprawling Long Beach plant. Barring congressional intervention or a spate of foreign orders—which analysts say is unlikely—the factory is expected to shut down completely by the end of next year.

of these industries and the importation of all computer and electronic products and communications equipment would be a national security nightmare. Specific policies based upon characteristics of these industries must be developed.

Figures 2.9a and 2.9b present the employment for the manufacturing of machinery. In this subsector, all of the industries have experienced declines, but two industries have experienced significantly smaller declines. The number of employees manufacturing agricultural, construction, and mining equipment declined only 7.1 percent from 1990–2012, and the number of employees manufacturing turbine and power transmission equipment declined by 10.3 percent. The other industries ranged from declines of 22.7 percent for HVAC and commercial refrigeration equipment to 33.6 percent for the manufacture of metalworking machinery. Since metalworking machinery is used to make other products ranging from airplanes to automobiles to appliances, the decline in employment of 33.6 percent in this industry is particularly troubling. In fact, two studies of the defense industrial base consider the metalworking machinery industry to be one of the most critical for national defense and innovation (Yudken 2010; Webber 2009).

Figures 2.10a and 2.10b present the employment in the manufacturing of the primary metals industries, and as may be seen, employment losses in these industries have been substantial (see Article 2.5). The

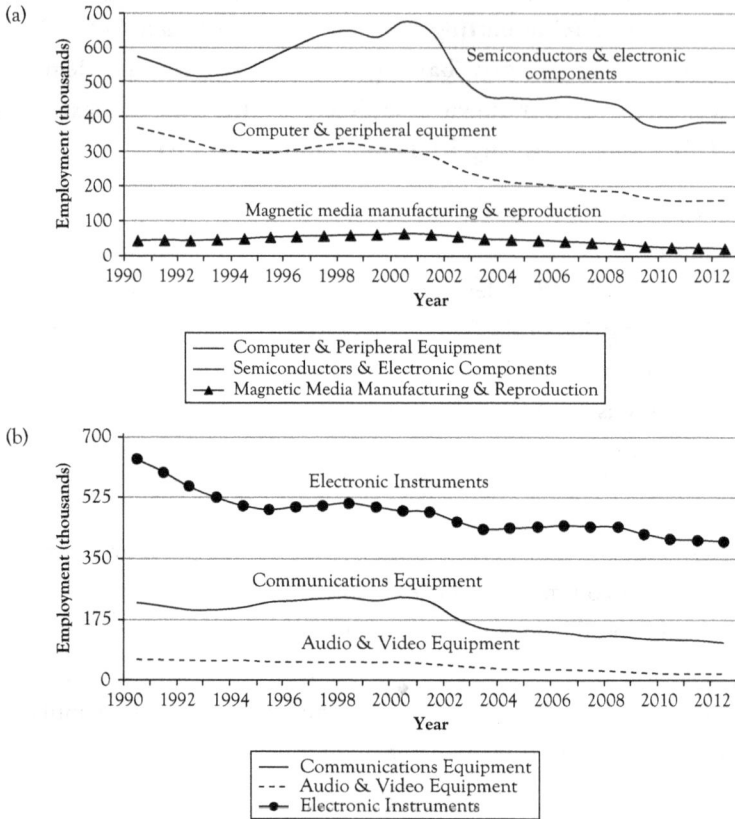

Figure 2.8 Durable goods—employment in computer and electronic products industries
Source: Bureau of Labor Statistics (2014a).

Article 2.4. Dell to close its Winston-Salem plant News-Record. com, Greensboro, N.C. November 20, 2010.

Dell's decision Wednesday to close its Forsyth County plant left 905 employees, business leaders and a community confounded by a reality that simply did not compute. Dell opened its desktop computer assembly factory just four years ago. More consumers want laptop computers, and the Winston-Salem plant churned out desktop computers for businesses. Dell chose not to assemble laptops here because it's less expensive to build them overseas. Production of servers—a computer system that provides essential services across a network—never materialized as hoped. Luring Dell was seen as a way to bring in a high-tech manufacturer who could replace job losses in industries such as textiles and furniture.

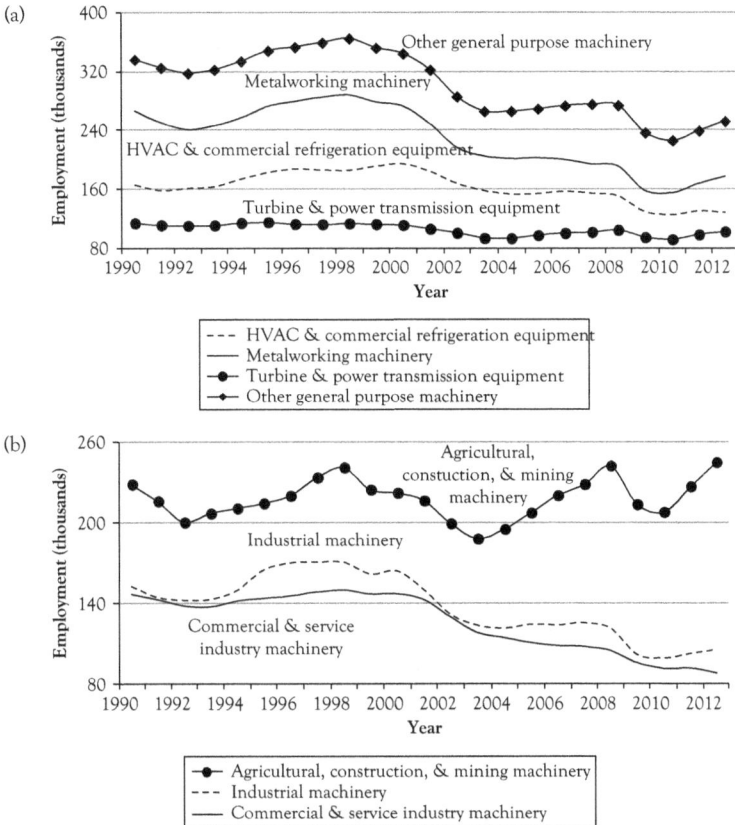

(a)

(b)

Figure 2.9 Durable goods—employment in machinery industries
Source: Bureau of Labor Statistics (2014a).

sharp decline in foundries is of particular concern. A foundry is a manufacturing facility that produces metal castings. Castings are widely used to manufacture parts in the automotive, aviation, and aerospace industries, and as such, foundries represent a critical supplier industry to these other industries. Webber listed foundries as a critical industry for defense (Webber 2009). Iron and steel mills and ferroalloy production also show a steep and on-going decline in employment.

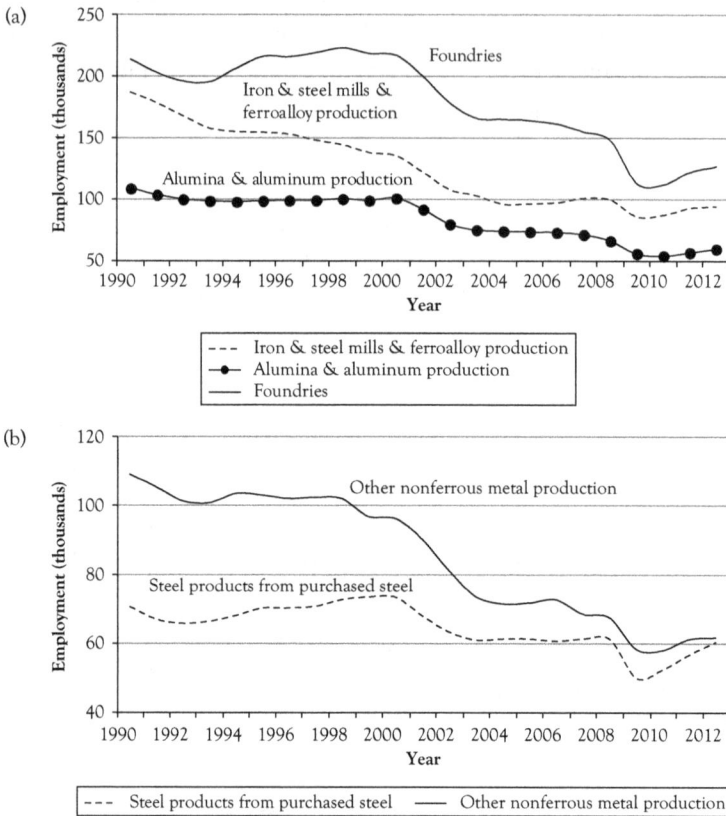

Figure 2.10 *Durable goods—employment in primary metals industries*

Source: Bureau of Labor Statistics (2014a).

Article 2.5. Wheland Foundry to close all operations by end of month; February 01, 2002, The Chattanoogan

Wheland Foundry officials said they will close all operations by the end of the month—throwing its remaining 1,100 workers out of jobs.

Wayne Tamme, Vice President of Human Resources, said the Middle Street plant just off South Broad "poured its last iron on Friday." That facility has some 400 workers.

Mr. Tamme said the Warrenton, Ga., plant will shut down in mid to late February. The main Broad Street plant and the centrifuge plant will close by the end of the month. The latter plants have some 700 workers.

Figures 2.11a, 2.11b, and 2.11c present the employment in fabricated metal products industries. These trend lines are cyclical with the recessions and do not exhibit such strong declines as in other industries. One important industry that does exhibit decline in employment, however, is the forging and stamping industry that lost 22.7 percent of employees between 1990 and 2012. The forging and stamping industry, similar to foundries, is critical because it underpins and supplies many other industries. Forging is a manufacturing process involving the shaping of metal using compressive forces—think blacksmiths in terms of an original form. Modern forges use complex presses or hammers powered by compressed air, electricity, hydraulics, or steam. Forged parts are widely used in motor vehicles, aviation, and machinery, and the loss of capacity in this industry is threatening to the United States.

Figure 2.12 presents employment trends for the industries in the electrical equipment and appliances subsector. Each industry shows sharp declines in the number of employees beginning around the year 2000. The overall subsector shed 42 percent of employees but the household appliance industry experienced a reduction of employment of 52 percent (see Articles 2.6 and 2.7).

Figure 2.13 presents employment history for the furniture and related products manufacturing subsector. As may be seen, the manufacturing industries in this area have been hard hit in terms of lost employment. Household and institutional furniture manufacturers lost roughly 215,000 jobs between 2000 and 2010. Office furniture and fixtures manufacturers lost approximately 80,000 over the same period. Again, a pattern may be observed where employment grew between 1990 and 2000 and then began a steep decline for both furniture manufacturing industries (see Article 2.8).

Summary

Summary Table 2.4 presents employment data for a number of manufacturing industries for the years 1990, 2000, 2007, 2009, and 2012. As may be seen, the period from 1990 to 2000 had mixed results for employment. On one hand, manufacturers of motor vehicle parts, forging and stamping, metalworking machinery, foundries, semiconductors and electronic components, communication equipment, household and institutional furniture, and office furniture all had *increases* in employment.

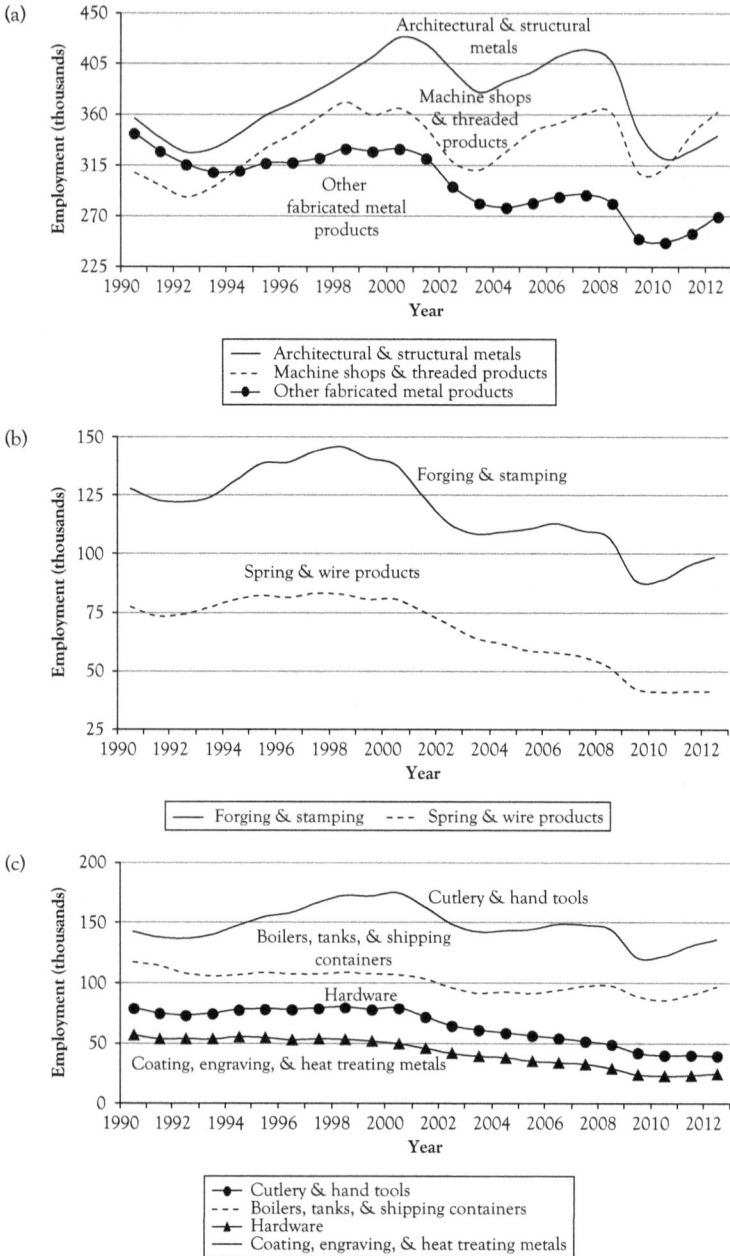

Figure 2.11 *Durable goods—employment in fabricated metal products industries*

Source: Bureau of Labor Statistics (2014a).

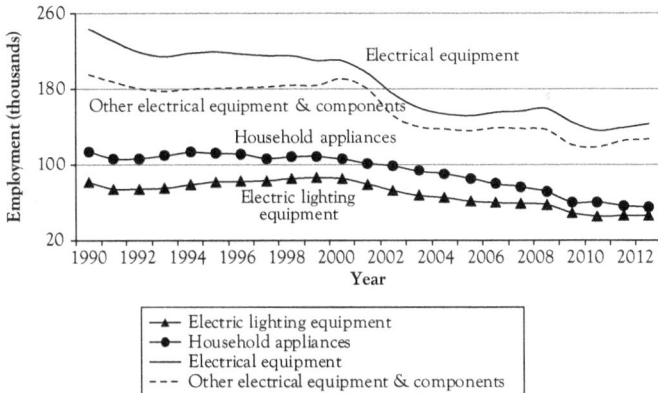

Figure 2.12 Durable goods—employment in electrical equipment, appliances, and components industries

Source: Bureau of Labor Statistics (2014a).

Article 2.6. Maytag closing Illinois plant by late 2004, idling 1,600; Deseret News, October 12, 2002

GALESBURG, Ill. (AP)—Maytag Corp. will close its refrigerator production plant in Galesburg by late 2004, affecting 1,600 workers, the company announced Friday.

The plant is no longer "competitively viable" in a refrigeration market that has seen competitors move production to Mexico over the past few years, Jim Little, Maytag's Vice President of Operations, wrote in a letter to employees.

Article 2.7. Historic Maytag factory shuts its doors; Manufacturing.net, October 10, 2007

NEWTON, Iowa (AP)—The last washers and dryers were rolling off Maytag assembly lines here Thursday, and the 550 workers who built them were preparing to leave the 2 million-square-foot factory for the last time, ending a century of appliance manufacturing in Newton under the iconic brand name.

For many workers, it's a sad parting with a company that has provided for their families over generations. Worries had mounted since
May 10, 2006—the day that plans were unveiled to close the Maytag
corporate headquarters and the local factory that made washers and
dryers, leaving about 1,800 local workers to find other jobs. At its
peak, Maytag had 4,000 workers in Newton, a town of 16,000 people
30 miles east of Des Moines.

**Article 2.8. Furniture plant closings hit Virginia region hard;
Thomas Russell, Furniture Today, May 24, 2010**

STANLEYTOWN, Va.—Stanley Furniture's plans to cease production of its adult case goods line here is the latest blow to a region that
has been hard hit by global competition and the recession. Stanley said
it will maintain a corporate headquarters and distribution center here,
employing about 140 people. It also will have an assembly and custom
finishing operation in nearby Martinsville, Va., that employs about 90.

Still, the plant closing will eliminate some 530 production jobs
in the last quarter of this year. Stanley will source its adult line from
plants in Southeast Asia, company officials said.

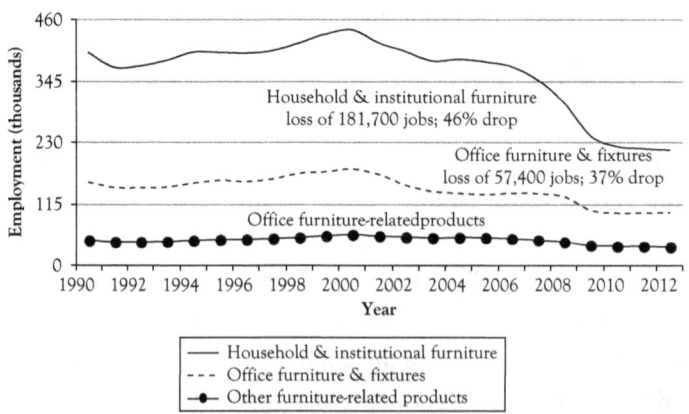

*Figure 2.13 Durable goods—employment in furniture and related
products industries*

Source: Bureau of Labor Statistics (2014a).

Another furniture manufacturer, American of Martinsville, also closed its Martinsville factory April 16, eliminating 224 jobs. That plant made Barcalounger products. These figures add to a string of job losses in the furniture industry in Virginia over the past 10 years. According to the Virginia Economic Development Partnership, 41 furniture plants have closed in Virginia since 2000, eliminating 7,237 jobs.

In Henry County, which is an hour north of High Point, six big plants have closed since 2000, costing 1,230 jobs. Among the companies that have closed plants are Bassett, Hooker, Pulaski, Ridgeway, and Vaughan

On the other hand, between 1990 and 2000, manufacturers of computer and peripheral equipment and household appliances had modest losses in employment and aerospace and cut and sew apparel had large losses. The period from 2000 to 2007, when most of the economy was growing, however, was very different for manufacturing industries, with large losses of manufacturing employment occurring in all industries except for a modest loss in aerospace. The onset of the Great Recession brought further job losses to all of these industries. Beginning in 2000, the fall-off in employment can be seen in nearly all of the trend graphs presented in this chapter. It is a remarkably similar pattern of losses.

Key Takeaways

With a few exceptions, most manufacturing industries had stable or growing employment in the nineties. Around the year 2000, however, manufacturing employment began a rapid decline across nearly all industries. Between 2000 and 2007 (before the recession), American manufacturing lost roughly four million jobs. The data pattern for most industries appears as:

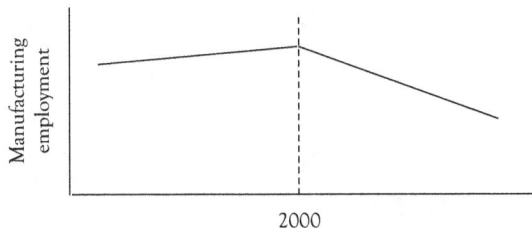

Data pattern 1: Employment

Table 2.4 Periods of growth and decline for selected industrial employment

	1990	2000	2007	2009	2012	% Change 1990–2000	% Change 2000–2007	% Change 2007–2009	% Change 2009–2012	% Change 1990–2012
Motor Vehicle Parts Employment (Thousands)	653.0	839.5	607.9	413.7	479.6	28.56	−27.59	−31.95	15.93	−26.55
Forging & Stamping Employment (Thousands)	128.1	138.2	110.0	89.0	99.0	7.88	−20.41	−19.09	11.24	−22.72
Metalworking Machinery Employment (Thousands)	266.7	273.5	193.9	157.6	177.1	2.55	−29.10	−18.72	12.37	−33.60
Foundries Employment (Thousands)	213.9	216.8	155.0	113.0	126.9	1.36	−28.51	−27.10	12.30	−40.67
Semiconductors & Electronic Components Employment (Thousands)	574.0	676.3	447.5	378.1	384.4	17.82	−33.83	−15.51	1.67	−33.03
Communications Equipment Employment (Thousands)	223.0	238.6	128.1	120.5	109.5	7.00	−46.31	−5.93	−9.13	−50.90
Computer & Peripheral Equipment Employment (Thousands)	367.4	301.9	186.2	166.4	158.6	−17.83	−38.32	−10.63	−4.69	−56.83

Aerospace Products & Parts										
Employment (Thousands)	840.7	516.7	489.2	492.2	497.4	-38.54	-5.32	0.61	1.06	-40.84
Household Appliances										
Employment (Thousands)	113.7	105.7	76.0	59.7	54.6	-7.04	-28.10	-21.45	-8.54	-51.98
Household & Institutional Furniture										
Employment (Thousands)	398.2	440.7	347.0	242.9	216.5	10.67	-21.26	-30.00	-10.87	-45.63
Office Furniture & Fixtures										
Employment (Thousands)	156.3	181.3	134.6	103.4	98.9	15.99	-25.76	-23.18	-4.35	-36.72
Cut & Sew Apparel										
Employment (Thousands)	749.6	380.2	165.6	132.0	122.1	-49.28	-56.44	-20.29	-7.50	-83.71
Total Manufacturing										
Employment (Thousands)	17,695	17,263	13,879	11,847	11,919	-2.44	-19.60	-14.64	0.61	-32.64

Source: Bureau of Labor Statistics (2014a).

CHAPTER 3

Loss of Establishments

Introduction

Some have argued that the loss of employment in manufacturing is tied to strong gains in productivity, that is, growth in the output per hour of labor. In other words, productivity gains arising from computers, robotics, and lean processes have replaced or reduced the number of employees needed to manufacture a product. In some instances, that argument may hold true, or, at least, be part of the story. On the other hand, however, general economic theory generally tends to argue that higher productivity leads to lower costs, lower prices, higher demand, and a resultant growth in employment, not a drop in employment (Nordhaus 2005). However, with the strong loss of employment occurring across just about all manufacturing industries, it is hard to argue and analytically defend the position that investments in productivity have led to growth in U.S. manufacturing employment. In fact, with large declines in manufacturing employment occurring across nearly all manufacturing industries, it is hard to imagine all of them making the required investments in productivity that would produce the significant drops in employment. Factors other than productivity must be at play.

It is argued here that a major factor in declining employment is the closing of factories; owners and corporations locking the doors and walking away. As Martin N. Baily, a chairman of the Council of Economic Advisors under President Clinton, said in a New York Times interview, "A lot of the reduction in employment is businesses deciding to close down operations or get out of a certain activity" (Goodman and Healy 2009). In these cases, laid-off employees have a particularly difficult time because their old job and plant are no longer there for them following a layoff. Closing a manufacturing establishment not only means no jobs for employees but it also means a loss of facilities and production infrastructure as well as salaries and wages and purchases of goods and services from other businesses flowing into the community. There is a negative ripple effect that moves out from a closed plant into the community.

Before presenting data on the number of establishments, it is important to understand the definition of "establishments." According to the BLS (Bureau of Labor Statistics 2014a), establishment is commonly understood as a single economic unit, such as a farm, a mine, a factory, or a store, that produces goods or services. Establishments are typically at one physical location and engaged in one, or predominantly one, type of economic activity for which a single industrial classification may be applied. A firm, or a company, is a business and may consist of one or more establishments, where each establishment may participate in different predominant economic activity. Data on establishments and employment by establishment are collected by BLS only in the first quarter of a calendar year.

Figures 3.1a and 3.1b present the total number of U.S. manufacturing establishments over the time period 1990–2012 (Bureau of Labor Statistics 2013). As may be seen in Figure 3.1a, the number of manufacturing establishments grew from 387,000 in 1990 to a peak of roughly 412,000 in 1998–99. Similar to what was seen with employment, the number of establishments then began a precipitous fall, losing roughly 50,000 manufacturing establishments by 2008. This pattern is highlighted in Figure 3.1b. With so many plants closing their doors, it is not difficult to imagine millions of jobs being lost. From 2008 to 2012 during the recession, another 17,000 manufacturing establishments were lost as were many more jobs. But again, "manufacturing" is an all-encompassing term that must be segmented to finer granularity to truly analyze the data. Figure 3.2 shows that the number of establishments involved in the manufacturing of durable goods actually increased by 20,000 between 1990 and 2000 and then began a sharp decline.

Figure 3.3 presents the number of establishments involved in the manufacture of nondurable goods. As may be seen, this number had very slight growth between 1990 and 2000 but then began a major decline.

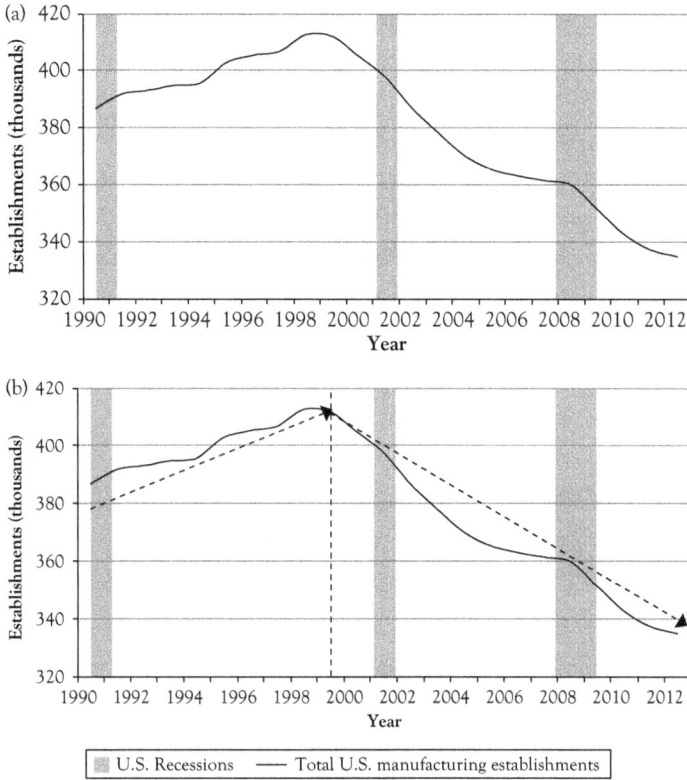

Figure 3.1 Total number of U.S. manufacturing establishments
Source: Bureau of Labor Statistics (2014).

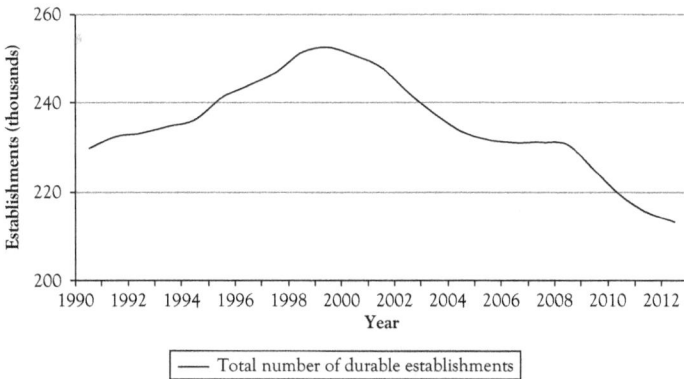

Figure 3.2 Total number of manufacturing establishments for durable goods
Source: Bureau of Labor Statistics (2014a).

Table 3.1 presents the change in the number of establishments for nondurable goods over the period 1990–2012. As may be seen, over 35,000 establishments that manufactured nondurable goods disappeared—representing a loss of 22.5 percent of these establishments. Several subsectors were particularly hard hit: textile mills (–50 percent), apparel (–65 percent), leather and allied products (–40 percent), and printing and related support activities (–38 percent). Other subsectors and industries showed relatively strong growth in the number of establishments including beverages (+64 percent) and chemicals (+20 percent), particularly pharmaceuticals and medicines (+75 percent). Nevertheless, the overall impact was a significant reduction in nondurable manufacturing establishments.

On the other hand, Table 3.2 shows that the loss of durable manufacturing establishments between 1990 and 2012 was not so severe, experiencing a loss of 16,572 establishments, a 7 percent decline. Half of this loss was in furniture and related products, a loss of 8,000 establishments, and other large losses were in the manufacturing of wood products and metalworking machinery. However, as seen in Figure 3.2, the number of durable goods manufacturing establishments grew strongly between 1990 and 2000 and then began a sharp decline and lost nearly 40,000 establishments between 2000 and 2012, a huge share of factories.

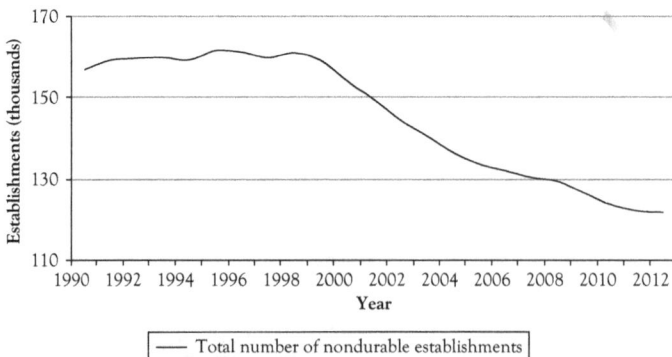

Figure 3.3 Total number of manufacturing establishments for nondurable goods

Source: Bureau of Labor Statistics (2014a).

Table 3.1 *Manufacturing of nondurable goods (Establishments and losses)*

NAICS	Sector Name	Establishments 1990	Establishments 2012	Establishments Loss	% Loss
	Nondurable Goods Total	156,926	121,644	−35,282	−22.48
311	**Food manufacturing**	**30,219**	**29,474**	**−745**	**−2.47**
3111	Animal food	2,157	2,042	−115	−5.33
3112	Grain and oilseed milling	901	909	8	0.89
3113	Sugar and confectionery products	2,582	1,938	−644	−24.94
3114	Fruit and vegetable preserving and specialty	2,076	1,934	−142	−6.84
3115	Dairy products	2,191	1,875	−316	−14.42
3116	Animal slaughtering and processing	5,044	3,998	−1,046	−20.74
3117	Seafood product preparation and packaging	1,293	825	−468	−36.19
3118	Bakeries and tortilla manufacturing	11,280	12,122	842	7.46
3119	Other food products	2,696	3,833	1,137	42.17
312	**Beverages and tobacco products**	**3,820**	**6,020**	**2,200**	**57.59**
3121	Beverages	3,573	5,866	2,293	64.18
3122	Tobacco and tobacco products	247	154	−93	−37.65
313	**Textile Mills**	**6,006**	**3,022**	**−2,984**	**−49.68**
3131	Fiber, yarn, and thread mills	632	407	−225	−35.60
3132	Fabric mills	2,188	1,113	−1,075	−49.13
3133	Textile and fabric finishing mills	3,186	1,502	−1,684	−52.86
314	**Textile product mills**	**8,681**	**7,128**	**−1,553**	**−17.89**
3141	Textile furnishings mills	4,366	2,161	−2,205	−50.50
3149	Other textile product mills	4,315	4,967	652	15.11
315	**Apparel**	**20,835**	**7,221**	**−13,614**	**−65.34**
3152	Cut and sew apparel	18,229	6,247	−11,982	−65.73
3151 & 3159	All other apparel manufacturing	1,319	660	−659	−49.96
316	**Leather and allied products**	**2,076**	**1,256**	**−820**	**−39.50**
322	Paper and paper products	6,544	5,652	−892	−13.63

(Continued)

Table 3.1 (Continued)

NAICS	Sector Name	Establishments 1990	Establishments 2012	Establishments Loss	% Loss
3221	Pulp, paper, and paperboard mills	769	799	30	3.90
3222	Converted paper products	5,775	4,853	−922	−15.97
323	**Printing and related support activities**	**48,452**	**30,271**	**−18,181**	**−37.52**
324	**Petroleum and coal products**	**2,261**	**2,372**	**111**	**4.91**
325	**Chemicals**	**13,473**	**16,190**	**2,717**	**20.17**
3251	Basic chemicals	2,179	2,998	819	37.59
3252	Resin, rubber, and artificial fibers	965	1,474	509	52.75
3253	Agricultural chemicals	1,189	1,187	−2	−0.17
3254	Pharmaceuticals and medicines	1,714	2,994	1,280	74.68
3255	Paints, coatings, and adhesives	2,166	1,917	−249	−11.50
3256	Soaps, cleaning compounds, and toiletries	2,602	2,721	119	4.57
3259	Other chemical products and preparations	2,659	2,900	241	9.06
326	**Plastics and rubber products**	**14,559**	**13,038**	**−1,521**	**−10.45**
3261	Plastics products	12,165	10,946	−1,219	−10.02
3262	Rubber products	2,395	2,092	−303	−12.65

Source: Bureau of Labor Statistics (2014a).

Table 3.2 Manufacturing of durable goods (establishments and losses)

NAICS	Sector Name	Establishments 1990	Establishments 2012	Establishments Loss	Loss
	Durable Goods Total	229,729	213,157	−16,572	−7.21
321	**Wood products**	**18,618**	**14,715**	**−3,903**	**−20.96**
3211	Sawmills & wood preservation	5,257	3,701	−1,556	−29.60
3212	Plywood & engineered wood products	1,809	1,642	−167	−9.23

Table 3.2 (Continued)

NAICS	Sector Name	Establishments 1990	Establishments 2012	Loss	% Loss
3219	Other wood products	11,552	9,373	−2,179	−18.86
327	**Nonmetallic mineral products**	**16,472**	**16,437**	**−35**	**−0.21**
3271	Clay products & refractories	2,451	1,562	−889	−36.27
3272	Glass & glass products	2,315	1,931	−384	−16.59
3273	Cement & concrete products	9,040	9,148	108	1.19
3274 & 3279	Lime, gypsum, & other nonmetallic mineral products	2,242	3,402	1,160	51.74
331	**Primary metals**	**5,963**	**5,615**	**−348**	**−5.84**
3311	Iron & steel mills & ferroalloy production	400	868	468	117.00
3312	Steel products from purchased steel	927	1,042	115	12.41
3313	Alumina & aluminum production	683	646	−37	−5.42
3314	Other nonferrous metal production	1,037	1,031	−6	−0.58
3315	Foundries	2,916	2,029	−887	−30.42
332	**Fabricated metal products**	**58,113**	**57,980**	**−133**	**−0.23**
3321	Forging & stamping	2,748	2,429	−319	−11.61
3322	Cutlery & hand tools	1,715	1,345	−370	−21.57
3323	Architectural & structural metals	13,666	14,399	733	5.36
3324	Boilers, tanks, & shipping containers	2,249	1,960	−289	−12.85
3325	Hardware	721	677	−44	−6.10
3326	Spring & wire products	1,889	1,379	−510	−27.00
3327	Machine shops and threaded products	21,841	22,905	1,064	4.87
3328	Coating, engraving, & heat treating metals	7,052	6,461	−591	−8.38
3329	Other fabricated metal products	6,232	6,427	195	3.13
333	**Machinery**	**33,626**	**28,991**	**−4,635**	**−13.78**
3331	Agricultural, construction, & mining machinery	3,486	3,905	419	12.02

(Continued)

Table 3.2 (Continued)

NAICS	Sector Name	Establishments 1990	Establishments 2012	Establishments Loss	% Loss
3332	Industrial machinery	4,297	3,748	−549	−12.78
3333	Commercial & service industry machinery	2,821	2,648	−173	−6.13
3334	HVAC & commercial refrigeration equipment	2,227	2,177	−50	−2.25
3335	Metalworking machinery	12,407	8,829	−3,578	−28.84
3336	Turbine & power transmission equipment	1,071	1,219	148	13.82
3339	Other general purpose machinery	7,318	6,466	−852	−11.64
334	**Computer and electronic products**	**18,920**	**18,761**	**−159**	**−0.84**
3341	Computer & peripheral equipment	1,938	1,616	−322	−16.62
3342	Communications equipment	2,292	2,282	−10	−0.44
3343	Audio & video equipment	659	659	0	0.00
3344	Semiconductors & electronic components	6,271	5,719	−552	−8.80
3345	Electronic instruments	6,459	7,594	1,135	17.57
3346	Magnetic media manufacturing & reproduction	1,301	891	−410	−31.51
335	**Electrical equipment and appliances**	**7,183**	**7,342**	**159**	**2.21**
3351	Electric lighting equipment	1,595	1,373	−222	−13.92
3352	Household appliances	439	472	33	7.52
3353	Electrical equipment	2,977	2,841	−136	−4.57
3359	Other electrical equipment & components	2,172	2,655	483	22.24
336	**Transportation equipment**	**14,229**	**14,232**	**3**	**0.02**
3361	Motor vehicles	375	470	95	25.33
3362	Motor vehicle bodies & trailers	2,546	1,998	−548	−21.52
3363	Motor vehicle parts	6,296	5,587	−709	−11.26
3364	Aerospace products & parts	2,559	3,102	543	21.22
3366	Ship & boat building	1,601	1,774	173	10.81

Table 3.2   (Continued)

NAICS	Sector Name	Establishments 1990	Establishments 2012	Establishments Loss	Establishments % Loss
3365 & 3369	Railroad rolling stock & other transportation equipment	598	977	379	63.38
337	**Furniture and related products**	**26,238**	**18,230**	**–8,008**	**–30.52**
3371	Household & institutional furniture	20,376	13,590	–6,786	–33.30
3372	Office furniture & fixtures	4,401	3,661	–740	–16.81
3379	Other furniture–related products	1,461	979	–482	–32.99
339	**Miscellaneous manufacturing**	**30,367**	**30,854**	**487**	**1.60**
3391	Medical equipment and supplies	12,905	12,978	73	0.57
3399	Other miscellaneous manufacturing	17,462	17,876	414	2.37

Source: Bureau of Labor Statistics (2014a).

Establishment Trends by Industries

To examine these declines in establishments more closely in terms of timing, Figures 3.4 through 3.11 present the number of establishments by
year for seven hard hit industries.

Figure 3.4 presents the number of establishments for cut and sew
apparel manufacturing, and Figure 3.5 presents the number of establishments for fabric mills. Both of these industries suffered devastating drops
in the number of manufacturing establishments (see Articles 3.1 and 3.2).
The losses of manufacturing establishments began around 1999 and 2000
and have continued unabated. In terms of strategy and potential government policy, as noted earlier, the jobs in apparel and textiles are almost
certainly gone for good with little hope for reshoring or rebuilding. It is
very hard to revive an industry after its sales, employment, and facilities
have declined so severely.

Similar declining trends in the number of establishments exist for
critical industries in the manufacture of durable goods. Figure 3.6 presents the number of establishments for foundries and for forging and

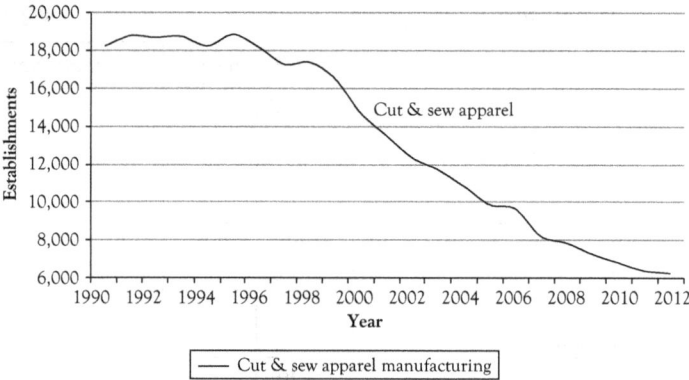

Figure 3.4 *Number of establishments for cut and sew apparel manufacturing*

Source: Bureau of Labor Statistics (2014a).

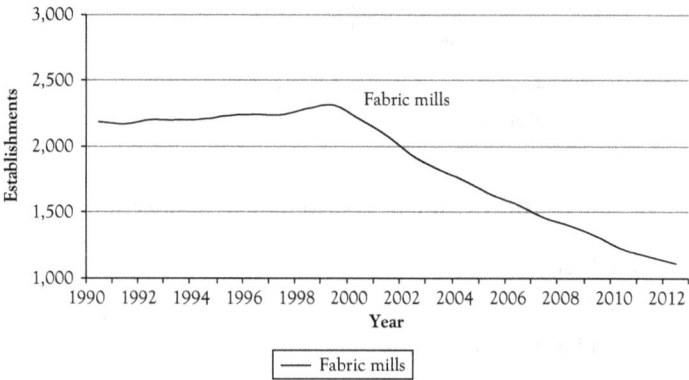

Figure 3.5 *Number of establishments for fabric mills*

Source: Bureau of Labor Statistics (2014a).

stamping. From 2000 to 2010, the number of foundries dropped from 2,916 to 2,128, losing roughly 800 facilities or 27.5 percent of the industry (see Articles 3.3, 3.4, and 3.5). As noted earlier, these two industries underpin many other industries. Loss of these capabilities is a threat to a robust manufacturing sector and to national defense. The Defense Production Act Committee formed a Metals Fabrication Study Group. They concluded "Domestic heavy forging capabilities are currently at risk because of market segmentation due to the low-volume, specialty demand of the Department of Defense (Office of the Deputy Assistant Secretary

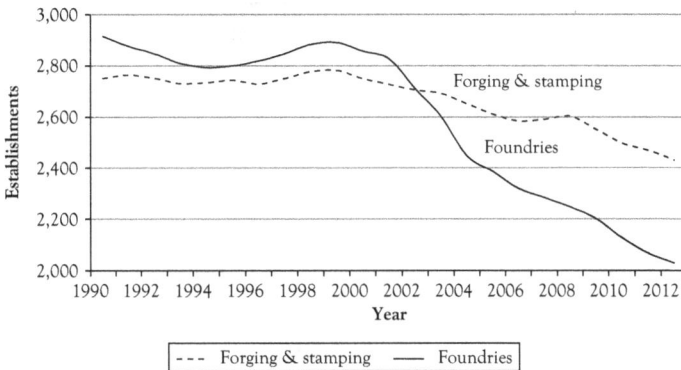

Figure 3.6 Number of establishments for forging and foundries
Source: Bureau of Labor Statistics (2014a).

Article 3.1. London Fog plant closing Rest of production now goes overseas to lands of low wages; Sean Somerville, Baltimore Sun, April 3, 1997

London Fog Industries Inc., which has manufactured raincoats in the city since 1922, said yesterday that it will close its Northwest Baltimore plant in June and eliminate 281 jobs. The closure of London Fog's sole remaining U.S. factory comes two years after the company reopened the shuttered facility with help from $1.8 million in state and city incentives and a $1.25-an-hour wage cut." The substantially lower costs of goods manufactured by overseas contractors, as compared to U.S.-based costs, has forced us, finally and reluctantly, to join our competitors in placing the remaining portion of our business with off-shore contractors," said Robert E. Gregory, chairman and chief executive of Eldersburg-based London Fog.

---

of Defense 2013). The companies that operate the forging presses face four primary challenges that threaten their ongoing viability: (1) aging forging infrastructure; (2) uncertain demand from the U.S. government; (3) undercapitalization; and (4) limited ingot supply that inhibits their ability to diversify to serve commercial customers." The Study Group recommended that the national defense needs in the areas of castings adaptability and machining be addressed in the near future.

Article 3.2. Where free trade hurts; Business Week, December 15, 2003

Under a 1974 global pact called the Multi-Fiber Arrangement (MFA), 47 nations each gets a share of the European and U.S. markets for clothing and textiles.

In 1995, the U.S. and Europe agreed to begin phasing out their quotas on clothing and textiles as part of the deal that created the World Trade Organization. Of the 140 categories of clothing covered by the MFA, quotas on about 50 less contentious categories have already been eliminated. By Jan. 1, 2005, the rest are scheduled to disappear, though most products will still face import duties of 16 percent in the U.S. and 18 percent in the European Union. Developing nations hailed the agreement. That was before China was invited to the party. In December, 2001, after 13 years of negotiations, China joined the WTO. Now, as a member of the global trading club, China will be able to compete on an equal footing to sew blazers, blouses, and bedspreads for the fashion-conscious consumers of Europe and America. The grand prize: $500 billion in global garment trade.

Suddenly, the much-maligned quota system looks like a lifeline. Rather than helping developing nations, the phase-out of quotas creates a Darwinian survival of the fittest—or, as critics of globalization would have it, a race to the bottom, where wages and benefits are certain to be sacrificed in a frantic effort to retain market share. When quotas on baby clothes and soft luggage ended last year, China's exports of baby clothes to the U.S. leaped 826 percent, and its soft luggage shipments rose fivefold. In Thailand, the Philippines, Indonesia, and Mexico, production of those products dropped by half. In the Dominican Republic, luggage exports plummeted by 70 percent, to $8.2 million. That kind of competition benefits consumers around the developed world. Prices have already fallen by 30 percent on dozens of items that went off quota last year, according to industry estimates. And buyers from companies such as the Gap and Nike have been flooding China in search of new suppliers in anticipation of the end of the quota regime.

Article 3.3. U.S. International Trade Commission, Foundry products: Competitive position in the U.S. Market, Investigation No. 332-460, USITC Publication 3771, May 2005

Many high-volume commodity-type castings are increasingly sourced from foreign suppliers.

Moreover, U.S. import statistics show increasing imports of downstream products containing the foundry product groups during 1999–2003, which suggests that purchases of downstream products containing foreign metal castings replaced use of U.S. foundry products.

In addition to quality, price is considered a significant factor in purchasing decisions, with nearly one-half of responding purchasers indicating that they usually buy castings at the lowest price. Although U.S. purchasers indicated that domestic castings producers have lowered prices, improved product quality, and shortened lead times to improve competitiveness, about one-third of reporting U.S. purchasers significantly increased their purchases of foreign castings at the expense of U.S. castings, primarily because of lower foreign pricing.

Article 3.4. American Foundry Society, Inc., 2001

Atchison Casting Corp. announced that it will close its PrimeCast, Inc. foundry unit, Beloit, Wisconsin, and South Beloit, Illinois. The company had previously announced a partial shutdown of the unit, following the closure of Beloit Paper Machinery Corp., which had been PrimeCast's major customer and the former owner of the foundry. The foundry was once South Beloit's second-largest employer and produced ductile, gray iron and stainless steel castings for nearly 43 years.

Article 3.5. U.S. Pipe and Foundry to close its North Birmingham plant, eliminating 260 jobs; Russell Hubbard—The Birmingham News; Thursday, February 04, 2010

U.S. Pipe and Foundry said Wednesday it plans to close the North Birmingham plant that has been a bedrock of the Collegeville neighborhood for more than 100 years, costing 260 jobs as metro-area unemployment stands at a 26-year high.

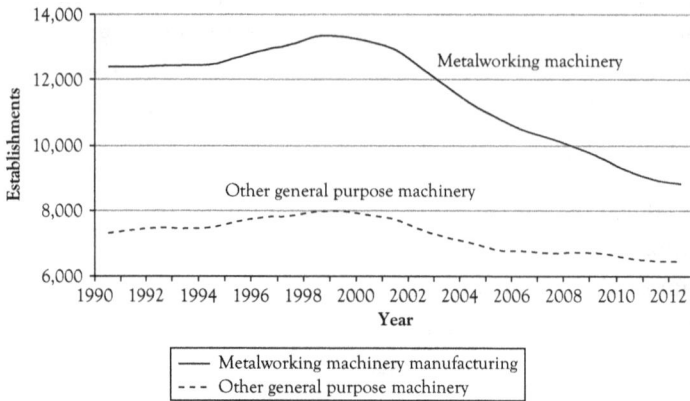

Figure 3.7 Number of establishments for metalworking
Source: Bureau of Labor Statistics (2014a).

Figure 3.7 presents yet another similar and disturbing decline in the number of establishments for the manufacture of metalworking machinery. Losing the ability to manufacture metalworking machinery again impacts many industries that require this equipment to manufacture their own products. Between 1999 and 2010, the number of establishments involved in the manufacture of metalworking machinery dropped from 13,321 to 9,209, a loss of roughly 4,200 facilities representing 31 percent of the industry.

Machine tools have long been considered essential to maintaining the country's national security. In 1948, Congress passed the National Industrial Reserve Act based on the idea that the "defense of the U.S. requires a national reserve of machine tools for the production of critical items of defense material." In 1986, President Ronald Reagan, a staunch free-trade advocate, supported a five-year Voluntary Restraint Agreement with Japan and Taiwan on imports of machine tools based on national-security grounds (Richman 1988; McCormack 2009). In making his determination, Reagan said the industry was a "vital component of the U.S. defense base."

Figure 3.8 presents the number of establishments manufacturing motor vehicle parts. This is one of the largest manufacturing industries in terms of employment, and it represents a vital aspect of manufacturing. As seen in Figure 3.8, the number of establishments grew from 6,296 in 1990 to 7,077 in 1999. The number of establishments manufacturing

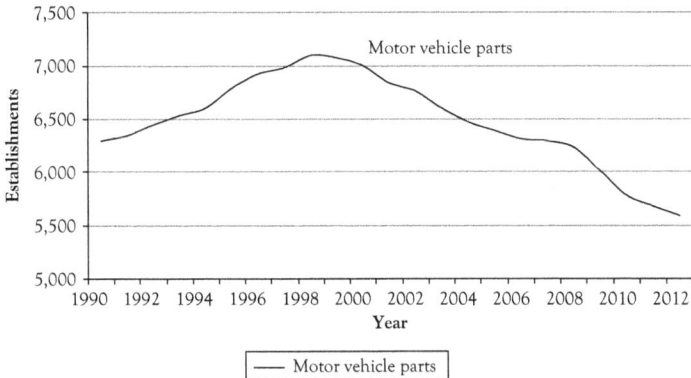

Figure 3.8 Number of establishments for motor vehicle parts
Source: Bureau of Labor Statistics (2014a).

motor vehicle parts then began a precipitous decline, dropping to 6,234 in 2008 and then to 5,780 in 2010, a total loss of 1,297 establishments.

An illustrative case is given by the closing of New United Motor Manufacturing, Inc. (NUMMI). On Thursday, April 1, 2010, the final Toyota Corolla left the NUMMI plant in Fremont, California, where General Motors (GM) and Toyota had engaged in a joint venture. GM ceased production at the plant in 2009, when the automaker was restructured. The closing by Toyota eliminated about 4,700 jobs (Dinkelspiel 2010). This closure then created ripples of closures. A report prepared for the California state treasurer estimated that NUMMI's closing would cost 20,000 jobs in businesses that supply the plant (Shaiken 2010). For example, InJex Industries, a supplier of door panels, laidoff most of its 380 employees the day before NUMMI closed (Dinkelspiel 2010). Undoubtedly, many other supplier plants that were dependent upon NUMMI also closed. U.S. motor vehicle parts supplier facilities have been closed and replaced by production in other countries (Alliance for American Manufacturing 2012).

Figure 3.9 presents the number of establishments manufacturing semiconductors and electronic components. This is one of the largest manufacturing industries in terms of employment, and it is a cornerstone of high tech manufacturing (see Article 3.6). As seen in Figure 3.9, the number of establishments grew from 6,271 in 1990 to 7,185 in 1999. The number of establishments manufacturing semiconductors and electronic

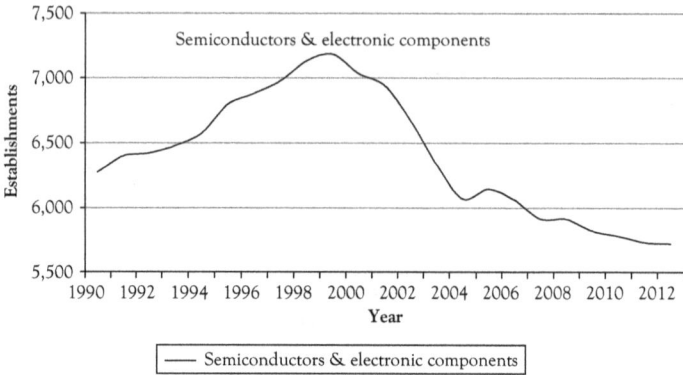

Figure 3.9 Number of establishments for semiconductors and electronic components

Source: Bureau of Labor Statistics (2014a).

Article 3.6. Layoffs and plant closings cap a bad year for Motorola; InsideChips.com, January 08, 2002

In mid-December, Motorola announced it not only expected to report in January its fourth consecutive quarterly loss, it was also laying off an additional 9,400 employees, closing an unspecified number of plants, and intended to eventually outsource 50 percent of its CMOS-based products. Prior to the announcement, Motorola had already given notice to 4,100 employees company-wide, and was in the process of notifying another 4,000 in the besieged Austin-based Semiconductor Products Sector that their jobs would end over the next year. Motorola said it is also eliminating about 1,300 positions in its equipment-manufacturing businesses. The company said shuttering the plants and reducing its workforce will save the company about $865 million in 2002 and about $1.1 billion in 2003. The newest layoffs will bring the total number of lost jobs to 48,400. The company's workforce is down one-third from its peak of 150,000 in August 2000.

---

components then began a rapid decline, dropping to 5,908 in 2008 and then to 5,777 in 2010, a total loss of 1,408 establishments.

As was seen in Figure 2.13, employment in the furniture industry has plunged dramatically (see Articles 3.7 and 3.8). The number

Article 3.7. Pulaski Furniture to close last U.S. plant; The company will still distribute imports, but it will no longer make domestic cabinets; Angela Manese-Lee; Roanoke Times, Saturday, February 17, 2007

Pulaski Furniture Corp. announced Friday that it will close its last domestic manufacturing plant by late April, laying off about 260 people and leveling a significant blow to Pulaski's economy. "This is part of a trend within the furniture industry that continues the shift of domestically manufactured furniture to off shore production, especially to Asia," Vice President of Operations Lamont Hope said in the release. "Our competition, by moving production off shore, has created such pricing pressure that our domestic operation can no longer compete."

Friday's announcement comes just four months after Pulaski Furniture announced in October that it would close another Pulaski plant, laying off 119 people in the process.

Founded in 1955, Pulaski Furniture has been a major employer in downtown Pulaski for decades.

Article 3.8. Kentwood mayor blasts Steelcase plant closing as company continues decade of local job cuts; Julia Bauer; The Grand Rapids Press; Published: Thursday, January 13, 2011, 9:38 AM; Updated: Thursday, January 13, 2011, 4:23 PM

GRAND RAPIDS—When Steelcase Inc. lays off or buys out 400 more of its West Michigan employees, it will be down to 3,000 workers here.

Ten years ago, the office furniture giant employed 9,000 in this region. The latest cutbacks will close the Kentwood East seating plant, along with a Vecta plant in Grand Prairie, Texas, and another in Markham, Ontario.

By next year, production of familiar makes of Steelcase chairs, including Think, Leap and the new node school chair, will move south of the border. A leased plant in Reynosa, Mexico, will make all Steelcase chairs for the North American market.

of manufacturing plants has also declined precipitously as shown in Figures 3.10 and 3.11. The states of Virginia and North Carolina have been particularly hard hit.

But the decline, and now end, of Pulaski Furniture's domestic manufacturing operations does not come as a surprise. In recent years, a number of communities in Southwest Virginia have struggled with furniture plant closings and mass layoffs, frequently blamed on foreign competition. And in the past 10 years, Pulaski Furniture has relied increasingly on imports, and now totally.

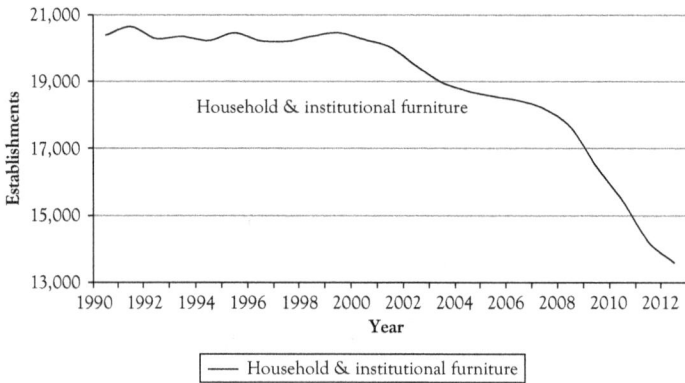

Figure 3.10 Number of establishments for household and institutional furniture

Source: Bureau of Labor Statistics (2014a).

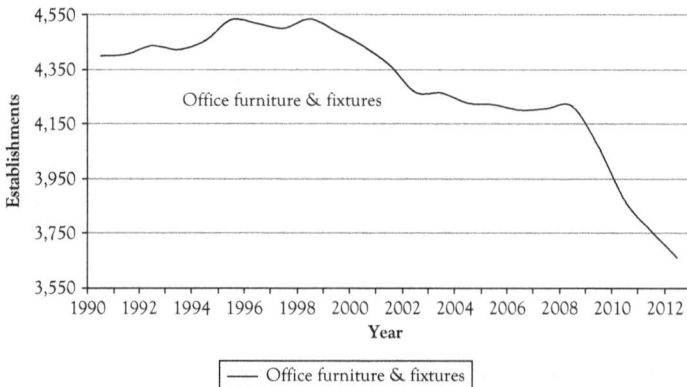

Figure 3.11 Number of establishments for office furniture and fixtures

Source: Bureau of Labor Statistics (2014a).

Summary

As was seen in Chapter 2 for the loss of employment, a very common pattern emerges for the loss of establishments. This pattern may be characterized somewhat generally by stability or slight growth or loss between 1990 and the year 2000 but then factories and plants began closing in nearly all industries. This pattern of an abrupt turn is not compatible with on-going productivity advances as seen in agriculture. Other forces are at play to cause such a dramatic turn in employment and establishment trend lines.

Key Takeaways

With a few exceptions, most manufacturing industries had stable or growing number of establishments in the nineties. Around the year 2000, however, the number of manufacturing establishments began a rapid decline across nearly all industries. Between 2000 and 2007 (before the recession), American manufacturing lost roughly 50,000 manufacturing facilities. The data pattern for most industries appears as:

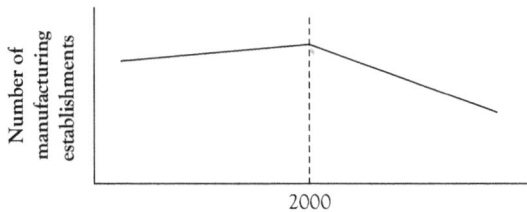

Data pattern 2: Establishments

CHAPTER 4

Observations on Employment and the Number of Establishments

Introduction

Previously presented figures and tables have shown that both manufacturing employment and the number of manufacturing establishments have seen substantial reductions since 1990. However patterns have emerged that differentiate certain periods within the time span between 1990 and 2012. The historical evolution tends to be segmented as follows:

1990	Mild recession
1991–1999/2000	Growth in employment; growth in the number of establishments Anomalies include aerospace and apparel which experienced declines in employment
2001	Mild recession
2001–2007	The economy is growing but it appears that a new trend is at work in manufacturing with significant reductions in the number of establishments and large losses in employment
2008–2009	Great recession, significant drops in establishments and losses in employment
2010–2012	Slow recovery from the recession

If one examines Figures 3.1a and 3.1b through 3.7 that present the number of establishments for various industries, a very similar pattern is seen. Starting around 2001, major declines begin to occur in the number of establishments. Two factors seem to be at play. First, the United States experienced a recession during 2001 which would account for some losses during that year. The second factor is that China became an official member of the World Trade Organization (WTO) on December 11, 2001,

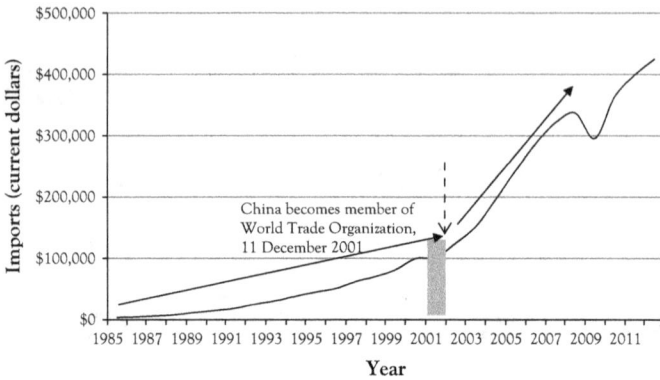

Figure 4.1 U.S. imports from China

after 15 years of negotiation. As seen in Figure 4.1, U.S. imports from China began to soar upwards immediately, tripling from $100 billion to over $300 billion in just six years. Although it is impossible to rigorously prove that the closure of plants occurred due to imports from China and the strategy of off-shoring, the timing certainly points to an interaction and relationship. The growth seems similar to that seen for apparel with the phase-out of quotas. These issues will be analyzed in some detail in Chapters 6 and 7.

Figure 4.2 from the Congressional Research Service annual report on auto parts shows the dramatic growth of imports of auto parts following 2001. This growth is mirrored by the decline in the number of establishments shown in Figure 3.8.

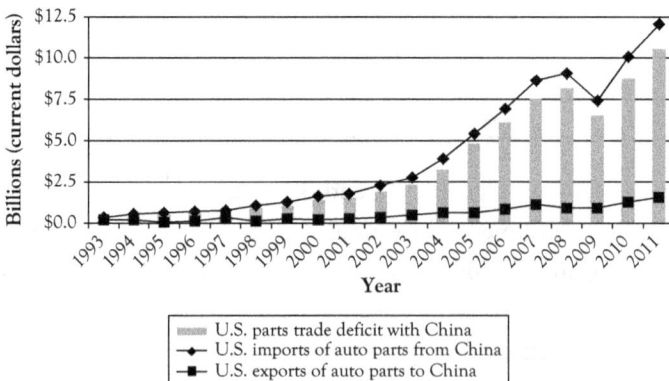

Figure 4.2 U.S.—China auto parts trade (1993–2011)

Source: (International Trade Administration 2011) International Trade Administration (http:// www.trade.gov/static/2011Parts.pdf)

Article 4.1. Costly trade with China; Robert E. Scott; Economic Policy Institute; October 9, 2007, Briefing Paper 188.

Contrary to the predictions of its supporters, China's entry into the World Trade Organization (WTO) has failed to reduce its trade surplus with the United States or increase overall U.S. employment. The rise in the U.S. trade deficit with China between 1997 and 2006 has displaced production that could have supported 2,166,000 U.S. jobs. Most of these jobs (1.8 million) have been lost since China entered the WTO in 2001. Between 1997 and 2001, growing trade deficits displaced an average of 101,000 jobs per year, or slightly more than the total employment in Manchester, New Hampshire. Since China entered the WTO in 2001, job losses increased to an average of 353,000 per year—more than the total employment in greater Akron, Ohio. Between 2001 and 2006, jobs were displaced in every state and the District of Columbia. Nearly three-quarters of the jobs displaced were in manufacturing industries. Simply put, the promised benefits of trade liberalization with China have been unfulfilled. Between 1997 and 2001, prior to China's entry into the WTO, the deficit increased $9 billion per year on average. Between 2001 and 2006, after China entered the WTO, the deficit increased $30 billion per year on average.

Article 4.2. Congressional Research Service Report for Congress: China's impact on the U.S. automotive industry; Stephen Cooney, Industry Specialist, Resources, Science, and Industry Division, Updated April 4, 2006

Chinese auto parts exports are already making inroads into the United States. While U.S. motor vehicle trade with China was insignificant in 2005, the United States imported $5.4 billion in parts from China, while it exported about one-tenth of that amount. China accounted for about 6 percent of U.S. auto parts imports in 2005, but the amount has quadrupled since 2000. Many of these imports are aimed at the aftermarket, as most of what China now exports to the U.S. market are standard products such as wheels, brake parts, and electronics. But with high rates of investment in China by the leading U.S. manufacturers of both cars and parts, major companies such as GM look to increase sourcing from China.

Table 4.1 presents employment, the number of establishments, and the average number of employees per establishment for selected industries and for total manufacturing for the years 1990, 2000, 2007, 2009, and 2012. As may be seen, with the exceptions of aerospace products and parts and cut and sew apparel, manufacturing industries experienced either growth or modest declines in employment during the period 1990–2000. During this period, the number of establishments tended to grow for manufacturing industries except for cut and sew apparel. For total manufacturing, between 1990 and 2000, employment declined by 2.4 percent and the number of establishments grew by 4.7 percent over this time period. The period from 2000–2007, however, exhibited vastly different dynamics.

During this period, every manufacturing industry in Table 4.1 experienced significant losses in employment. Moreover, every industry except aerospace experienced a sizeable reduction in the number of establishments during this period. In summary, for all manufacturing, between 2000 and 2007, roughly three and a half million manufacturing jobs were lost and 44,000 manufacturing facilities were closed. And all of these declines occurred before the onset of the Great Recession. During the Great Recession of 2007–2009, every industry in Table 4.1 then experienced additional losses in employment and the number of establishments.

Summary

An important pattern emerges in the data of Table 4.1. Every manufacturing industry in the table experienced a moderate or sharp decline in the average number of employees per establishment. Between 1990 and 2010, total manufacturing employment declined by 34.9 percent and the

Table 4.1 Periods of growth and decline in employment and the number of establishments for selected durable goods industries

	1990	2000	2007	2009	2012	% Change 1990–2000	% Change 2000–2007	% Change 2007–2009	% Change 2009–2012	% Change 1990–2007
Motor Vehicle Parts										
Employment (Thousands)	653.0	839.5	607.9	413.7	479.6	28.56	-27.59	-31.95	15.93	-26.55
Establishments	6,296	7,003	6,296	6,013	5,587	11.23	-10.10	-4.49	-7.08	-11.26
Employment Per Establishment	103.7	119.9	96.6	68.8	85.8	15.58	-19.46	-28.74	24.77	-17.23
Forging & Stamping										
Employment (Thousands)	128.1	138.2	110.0	89.0	99.0	7.88	-20.41	-19.09	11.24	-22.72
Establishments	2,748	2,748	2,589	2,550	2,429	0.00	-5.79	-1.51	-4.75	-11.61
Employment Per Establishment	46.6	50.3	42.5	34.9	40.8	7.88	-15.52	-17.85	16.78	-12.57
Metalworking Machinery										
Employment (Thousands)	266.7	273.5	193.9	157.6	177.1	2.55	-29.10	-18.72	12.37	-33.60
Establishments	12,407	13,169	10,217	9,605	8,829	6.14	-22.42	-5.99	-8.08	-28.84
Employment Per Establishment	21.5	20.8	19.0	16.4	20.1	-3.38	-8.62	-13.54	22.25	-6.69
Foundries										
Employment (Thousands)	213.9	216.8	155.0	113.0	126.9	1.36	-28.51	-27.10	12.30	-40.67
Establishments	2,916	2,858	2,285	2,204	2,029	-1.99	-20.05	-3.54	-7.94	-30.42
Employment Per Establishment	73.4	75.9	67.8	51.3	62.5	3.41	-10.58	-24.42	21.99	-14.74

(Continued)

Table 4.1 (Continued)

	1990	2000	2007	2009	2012	% Change 1990–2000	% Change 2000–2007	% Change 2007–2009	% Change 2009–2012	% Change 1990–2007
Semiconductors & Electronic Components										
Employment (Thousands)	574.0	676.3	447.5	378.1	384.4	17.82	-33.83	-15.51	1.67	-33.03
Establishments	6.271	7,035	5,911	5,817	5,719	12.18	-15.98	-1.59	-1.68	-8.80
Employment Per Establishment	91.5	96.1	75.7	65.0	67.2	5.03	-21.25	-14.14	3.41	-26.57
Communications Equipment										
Employment (Thousands)	223.0	238.6	128.1	120.5	109.5	7.00	-46.31	-5.93	-9.13	-50.90
Establishments	2,292	2,968	2,187	2,210	2,282	29.49	-26.31	1.05	3.26	-0.44
Employm ent Per Establishm ent	97.3	80.4	58.6	54.5	48.0	-17.37	-27.14	-6.91	-12.00	-50.68
Computer & Peripheral Equipment										
Employment (Thousands)	367.4	301.9	186.2	166.4	158.6	-17.83	-38.32	-10.63	-4.69	-56.83
Establishments	1,938	2,320	1,709	1,671	1,616	19.71	-26.34	-2.22	-3.29	-16.62
Employm ent Per Establishment	189.6	130.1	109.0	99.6	98.1	-31.36	-16.27	-8.60	-1.44	-48.23
Aerospace Products & Parts										
Employm ent (Thousands)	840.7	516.7	89.2	492.2	497.4	-38.54	-5.32	0.61	1.06	-40.84
Establishments	2,559	2,893	2,997	3,083	3,102	13.05	3.59	2.87	0.62	21.22
Employm ent Per Establishment	328.5	178.6	163.2	159.6	160.3	-45.64	-8.61	-2.19	0.44	-51.19

Table 4.1 (Continued)

	1990	2000	2007	2009	2012	% Change 1990–2000	% Change 2000–2007	% Change 2007–2009	% Change 2009–2012	% Change 1990–2007
Household Appliances										
Employment (Thousands)	113.7	105.7	76.0	59.7	54.6	-7.04	-28.10	-21.45	-8.54	-51.98
Establishments	439	546	522	508	472	24.37	-4.40	-2.68	-7.09	7.52
Employment Per Establishment	259.0	193.6	145.6	117.5	115.7	-25.25	-24.79	-19.28	-1.57	-55.34
Household & Institutional Furniture										
Employment (Thousands)	398.2	440.7	347.0	242.9	216.5	10.67	-21.26	-30.00	-10.87	-45.63
Establishments	20,376	20,253	18,175	16,471	13,590	-0.60	-10.26	-9.38	-17.49	-33.30
Employment Per Establishment	19.5	21.8	19.1	14.7	15.9	11.35	-12.26	-22.76	8.03	-18.48
Office Furniture & Fixtures										
Employment (Thousands)	156.3	181.3	134.6	103.4	98.9	15.99	-25.76	-23.18	-4.35	-36.72
Establishments	4,401	4,438	4,207	4,069	3,661	0.84	-5.21	-3.28	-10.03	-16.81
Employment Per Establishment	35.5	40.9	32.0	25.4	27.0	15.03	-21.68	-20.57	6.31	-23.93
Total Manufacturing										
Employment (Thousands)	17,695	17,263	13,879	11,847	11,919	-2.44	-19.60	-14.64	0.61	-32.64
Establishments (Thousands)	387	405	361	351	335	4.68	-10.71	-2.80	-4.70	-13.41
Employment Per Establishment	45.8	42.7	38.4	33.7	35.6	-6.81	-9.96	-12.18	5.57	-22.21

Source: Bureau of Labor Statistics (2014a).

Table 4.2 *Periods of growth and decline in employment and the number of establishments for selected nondurable goods industries*

	1990	2000	2007	2009	2012	% Change 1990–2000	% Change 2000–2007	% Change 2007–2009	% Change 2009–2012	% Change 1990–2007
Textile Mills										
Employment (Thousands)	491.8	378.2	169.7	124.4	118.0	-23.10	-55.13	-26.69	-5.14	-76.01
Establishments	6,006	6,027	3,828	3,463	3,022	0.35	-36.49	-9.54	-12.73	-49.68
Employment Per Establishment	81.9	62.8	44.3	35.9	39.0	-23.37	-29.35	-18.97	8.70	-52.31
Textile Product Mills										
Employment (Thousands)	235.6	229.6	157.7	125.7	116.6	-2.55	-31.32	-20.29	-7.24	-50.51
Establishments	8,681	8,710	8,130	7,810	7,128	0.33	-6.66	-3.94	-8.73	-17.89
Employment Per Establishment	27.1	26.4	19.4	16.1	16.4	-2.87	-26.42	-17.03	1.64	-39.73
Apparel										
Employment (Thousands)	902.8	483.5	214.6	167.5	148.1	-46.44	-55.62	-21.95	-11.58	-83.60
Establishments	20,835	16,816	9,492	8,339	7,221	-19.29	-43.55	-12.15	-13.41	-65.34
Employment Per Establishment	43.3	28.8	22.6	20.1	20.5	-33.64	-21.37	-11.16	2.11	-52.67
Paper & Paper Products										
Employment (Thousands)	647.2	604.7	458.2	407.0	379.0	-6.57	-24.23	-11.17	-6.88	-41.44
Establishments	6,544	7,049	6,315	6,144	5,652	7.72	-10.41	-2.71	-8.01	-13.63
Employment Per Establishment	98.9	85.8	72.6	66.2	67.1	-13.26	-15.42	-8.70	1.23	-32.20
Printing & Related Support Activities										
Employment (Thousands)	808.5	806.8	622.0	521.9	462.1	-0.21	-22.91	-16.09	-11.46	-42.84
Establishments	48,452	44,117	35,988	33,796	30,271	-8.95	-18.43	-6.09	-10.43	-37.52
Employment Per Establishment	16.7	18.3	17.3	15.4	15.3	9.60	-5.49	-10.65	-1.15	-8.52

Source: Bureau of Labor Statistics (2014a).

number of manufacturing establishments declined by a lesser amount of 11.4 percent. As a result, the average establishment had fewer employees. For example, in the manufacturing of motor vehicle parts, the average number of employees per establishment dropped from 104 to 72 in the years between 1990 and 2010. For foundries, the decline was from 73 to 52 employees per establishment. For total manufacturing, the drop was from 46 to 34 employees per establishment. This reduction in average size can occur in two separate fashions or a combination of the two ways. The first is simply that all establishments remaining in 2010 reduced their employee head count proportionately, the mix between large and small establishments remaining the same. The second mechanism by which the average might drop would arise from a different mix in the size of establishments, that is, fewer large establishments and relatively more small establishments. This would occur if the closing facilities were predominately those with a larger number of employees.

This would appear to be the case from examining the data in Table 4.1. For example, for motor vehicle parts manufacturing between 1990 and 2010, employment dropped by 234,000, while at the same time, 516 establishments were lost. This implies a ratio of 454 lost jobs per lost establishment. This is considerably larger than the average establishment size of 72 in 2010. For the forging and stamping industry, 39,100 employees were lost between 1990 and 2010, while 252 establishments were lost. In this industry, the ratio of lost jobs to lost facilities was 155 employees per establishment, considerably larger than the 2010 average of 36 employees per establishment. The figure is even more striking for the computers and peripheral equipment industry. In this industry, employment loss was 209,800, while the loss of establishments was 281. The lost employment per lost establishment in the computer and peripheral equipment industry was 747, much larger than the average of 95 employees per establishment in 2010. For the overall manufacturing industry, the lost employment was 6,167,000, while the number of lost establishments was 44,000, yielding 140 lost employees per lost establishment. Again, this figure is much larger than the 2010 average of 34 employees per establishment. This phenomenon will be examined for multiple industries in detail in the next chapter.

Key Takeaways

For most industries, the percentage of employment loss was much greater than the percentage drop in the number of establishments. This data pattern points to the loss of larger establishments—those establishments more economical and profitable to send off-shore.

CHAPTER 5

Manufacturing Plant Closings by Size

Introduction

Additional data analysis shows even more clearly that the United States has lost a greater number of larger manufacturing facilities than smaller ones. All establishment data are from the Bureau of Labor Statistics (BLS) website (Bureau of Labor Statistics 2013).

The BLS records the size of establishments as follows:

Size Code 1	Number of Employees	1–4
Size Code 2	Number of Employees	5–9
Size Code 3	Number of Employees	10–19
Size Code 4	Number of Employees	20–49
Size Code 5	Number of Employees	50–99
Size Code 6	Number of Employees	100–249
Size Code 7	Number of Employees	250–499
Size Code 8	Number of Employees	500–999
Size Code 9	Number of Employees	1,000 or more

The change over time in the mix of establishments by the nine size codes reveals how the structure of the various industries has changed since 1990. In general, it will be seen that every industry has become more concentrated in small to medium size plants with closings dominantly hitting large facilities.

Establishment Closings in the Manufacture of Durable Goods

Table 5.1a presents for each size code of the motor vehicle parts manufacturing industry the number of establishments in that size code, the number of employees working at those establishments, and the average number of employees for the establishments in that size code, for the years 1990, 1995, 2000, 2005, 2008, 2010, and 2012. Between 1990 and 2000, it can be seen that the number of plants increased strongly for the four largest size codes, 6 through 9. Note in Table 5.1a that for size code 9, 1,000 or more employees, the number of establishments dropped from 113 to 63 between 1990 and 2008. (Recall that establishment data are from the first quarter of the year so 2008 would be right at the beginning of the recession.) Then the number of size code 9 establishments fell to only 36 in 2010. The number then rebounded somewhat to 52 in 2012. The sharp decline between 2008 and 2010 and then the rebound most likely indicate a combination of plant closings and layoffs since the rebound in the number of establishments is most likely the result of workers being rehired and increasing the employment back over 1,000. This pattern is also seen for size codes 7 and 8 although these two size codes did not experience quite such large drops. The loss of facilities and motor vehicle parts manufacturing plants clearly occurred primarily at the largest facilities. Total employment for size code 9 dropped from 286,806 in 1990 to 79,855 in 2012, a staggering loss of 206,951 jobs in this size code alone.

Table 5.1a clearly shows that the primary loss of establishments occurred within the largest size group. But what about employment—in which size group did the largest employment drop occur? Let ε be the number of establishments for a certain size code in 1990 and $E + \delta E$ be the number of establishments in 2012 for that same size code. Let ε be the number of employees per establishment for that same size code in 1990 and $\varepsilon + \delta\varepsilon$ be the number of employees per establishment for that size code in 2012. Then total employment for the size code in 1990 is given by $(E)(\varepsilon)$ and the total employment for that size code in 2012 is given by $(E + \delta E)(\varepsilon + \delta\varepsilon)$. The change in employment between 1990 and 2012 is then given as,

$$\Delta \text{Employment} = (E + \delta E)(\varepsilon + \delta\varepsilon) - (E)(\varepsilon)$$

Which then yields:

$$\Delta \text{Employment} = (\varepsilon)(\delta E) + (E)(\delta\varepsilon) + (\delta\varepsilon)(\delta E)$$

Table 5.1a Distribution of establishment size for motor vehicle parts industry

Year	Size Code 1 <5 # of Estab.	Employees	Emp/Est	Size Code 2 5–9 # of Estab.	Employees	Emp/Est	Size Code 3 10–19 # of Estab.	Employees	Emp/Est
1990	1,655	3,239	2.0	850	5,666	6.7	886	12,217	13.8
1995	1,819	3,647	2.0	918	6,095	6.6	826	11,343	13.7
2000	2,145	3,638	1.7	860	5,745	6.7	772	10,740	13.9
2005	1,938	3,395	1.8	764	5,012	6.6	729	9,987	13.7
2008	2,009	4,275	2.1	777	5,255	6.8	674	9,481	14.1
2010	2,068	3,428	1.7	732	4,891	6.7	631	8,832	14.0
2012	1,871	3,216	1.7	708	4,666	6.6	618	8,679	14.0

Year	Size Code 4 20–49 # of Estab.	Employees	Emp/Est	Size Code 5 50–99 # of Estab.	Employees	Emp/Est	Size Code 6 100–249 # of Estab.	Employees	Emp/Est
1990	980	31,024	31.7	605	42,885	70.9	710	114,325	161.0
1995	971	30,629	31.5	618	43,860	71.0	764	124,793	163.3
2000	969	30,665	31.6	598	42,707	71.4	786	127,955	162.8

(Continued)

Table 5.1a (Continued)

Year	Size Code 4 20-49 # of Estab.	Employees	Emp/Est	Size Code 5 50-99 # of Estab.	Employees	Emp/Est	Size Code 6 100-249 # of Estab.	Employees	Emp/Est
2005	819	26,299	32.1	612	43,562	71.2	799	132,079	165.3
2008	803	25,517	31.8	618	44,441	71.9	763	125,593	164.6
2010	791	25,385	32.1	526	37,597	71.5	647	102,898	159.0
2012	725	23,343	32.2	504	35,419	70.3	643	102,699	159.7

Year	Size Code 7 250-499 # of Estab.	Employees	Emp/Est	Size Code 8 500-999 # of Estab.	Employees	Emp/Est	Size Code 9 1000 or more # of Estab.	Employees	Emp/Est
1990	346	119,390	345.1	151	103,404	684.8	113	286,806	2,538.1
1995	451	159,260	353.1	217	149,025	686.8	128	280,266	2,189.6
2000	568	197,132	347.1	234	160,253	684.8	117	264,977	2,264.8
2005	461	161,299	349.9	203	141,014	694.7	87	165,308	1,900.1
2008	395	136,270	345.0	177	120,795	682.5	63	102,547	1,627.7
2010	274	93,682	341.9	109	74,037	679.2	36	54,878	1,524.4
2012	337	117,577	348.9	143	94,364	659.9	52	79,855	1,535.7

Source: Bureau of Labor Statistics (2013).

Table 5.1b Motor vehicle parts industry change in employment 1990–2012

Size Code	Change in Employment Due to Change in Establishments			Size Code	Change in Employment Due to Change in Emp/Est			Size Code	Change in Employment Due to Combined Effect			Size Code	Change in Total Employment
	(Δ Est)	(EPE.)	=		(Δ EPE)	(Est.)	=		(Δ Est)	(Δ EPE)	=		
1	216	-2.0	423	1	-0.2	1,655	-395	1	216	-0.2	-52	1	-24
2	-142	6.7	-946	2	-0.1	850	-63	2	-142	-0.1	11	2	-999
3	-268	13.8	-3,695	3	0.3	886	227	3	-268	0.3	-69	3	-3,537
4	-255	31.7	-8,072	4	0.5	980	530	4	-255	0.5	-138	4	-7,680
5	-101	70.9	-7,159	5	-0.6	605	-368	5	-101	-0.6	62	5	-7,466
6	-67	161.0	-10,788	6	-1.3	710	-925	6	-67	-1.3	87	6	-11,626
7	-9	345.1	-3,106	7	3.8	346	1,327	7	-9	3.8	-35	7	-1,813
8	-8	684.8	-5,478	8	-24.9	151	-3,761	8	-8	-24.9	199	8	-9,040
9	-61	2,538.1	-154,824	9	-1,002.4	113	-113,275	9	-61	-1,002.4	61,149	9	-206,951

TOTAL LOSS OF EMPLOYMENT   -249,137

The first term in the equation depends upon the change in the number of establishments, δE, the second term depends upon the change in the average number of employees per establishment, $\delta \varepsilon$, and the third term is a combination of the changes and might be deemed a second order effect.

For size code 9 in the motor vehicle parts industry, the initial number of establishments ε in 1990 was 132, the initial number of employees per establishment ε was 2,538, the change in the number of establishments between 1990 and 2012, δE, was –61, that is a loss of 61, and the change in the number of employees per establishment was –1,002, that is a reduction of 1,002 in the number of employees per plant for size code 9. As shown in Table 5.1b, the change in employment due to the change in the number of establishments was –154,824; the change in employment from the change in the average number of employees per establishment was –113,275; and the change in employment arising from the third term in the equation was +61,149. The total change in employment for size code 9 was then –206,951, a loss of 206,951. As may be seen in Table 5.1b, the total change in employment between 1990 and 2012 for the motor vehicle parts industry was a loss of 249,137 and of this total loss, 206,951 came from size code 9 alone.

Table 5.2a presents for each size code of the foundries industry the number of establishments in that size code, the number of employees working at those establishments, and the average number of employees for the establishments in that size code, for the years 1990, 1995, 2000, 2005, 2008, 2010, and 2012. Note in Table 5.2a that for size code 9, with 1,000 or more employees, the number of establishments dropped from 21 to only 4 between 1990 and 2010, a reduction of 81 percent. Following the recession the number of size code 9 facilities increased from 4 to 5, again probably because of workers being recalled rather than the building of a huge new facility which is very unlikely. For the next largest size code, 8, having between 500 and 999 employees, the number of establishments dropped from 47 to 20, a loss of 57 percent. This size code then had a rebound to 29 establishments, again most likely due to workers being recalled and increasing employment. For this industry, the loss of facilities and manufacturing plants occurred in every size code but the largest loss of employment was in the larger facilities.

Table 5.2a Distribution of establishment size for foundries industry

	Size Code 1 < 5			Size Code 2 5–9			Size Code 3 10–19		
Year	# of Estab.	Employees	Emp/Est	# of Estab.	Employees	Emp/Est	# of Estab.	Employees	Emp/Est
1990	532	1,199	2.3	354	2,448	6.9	446	6,267	14.1
1995	503	976	1.9	355	2,432	6.9	421	5,894	14.0
2000	590	1,011	1.7	353	2,410	6.8	408	5,701	14.0
2005	552	1,021	1.8	316	2,154	6.8	338	4,796	14.2
2008	509	1,049	2.1	301	2,023	6.7	308	4,309	14.0
2010	590	1,963	3.3	293	1,978	6.8	332	4,794	14.4
2012	494	857	1.7	269	1,822	6.8	304	4,234	13.9

	Size Code 4 20–49			Size Code 5 50–99			Size Code 6 100–249		
Year	# of Estab.	Employees	Emp/Est	# of Estab.	Employees	Emp/Est	# of Estab.	Employees	Emp/Est
1990	686	21,295	31.0	379	26,903	71.0	357	55,775	156.2
1995	620	19,528	31.5	359	25,681	71.5	352	55,073	156.5
2000	605	19,732	32.6	364	25,535	70.2	335	52,803	157.6

(Continued)

Table 5.2a (Continued)

Year	Size Code 4 20–49			Size Code 5 50–99			Size Code 6 100–249		
	# of Estab.	Employees	Emp/Est	# of Estab.	Employees	Emp/Est	# of Estab.	Employees	Emp/Est
2005	489	15,790	32.3	289	20,013	69.2	263	40,471	153.9
2008	450	14,396	32.0	287	20,071	69.9	254	38,501	151.6
2010	420	13,346	31.8	240	16,558	69.0	186	27,774	149.3
2012	389	12,634	32.5	257	18,231	70.9	204	31,716	155.5

Year	Size Code 7 250–499			Size Code 8 500–999			Size Code 9 1000 or more		
	# of Estab.	Employees	Emp/Est	# of Estab.	Employees	Emp/Est	# of Estab.	Employees	Emp/Est
1990	108	36,631	339.2	47	31,660	673.6	21	33,457	1,593.2
1995	126	42,618	338.2	59	38,822	658.0	14	25,366	1,811.9
2000	140	47,910	342.2	57	37,964	666.0	17	25,773	1,516.1
2005	111	38,284	344.9	45	30,614	680.3	8	13,824	1,728.0
2008	100	34,512	345.1	37	25,045	676.9	8	11,639	1,454.9
2010	63	21,290	337.9	20	12,681	634.0	4	5,946	1,486.6
2012	87	30,047	345.4	29	20,128	694.1	5	7,563	1,512.5

Source: Bureau of Labor Statistics (2013).

Table 5.2b presents the change in employment for each size code between 1990 and 2012 for the foundries industry. For this industry, over half of the total loss of employment occurred in just two size codes, 6 and 9. Moreover, essentially all of the total loss derived from the loss of establishments rather than declines in the average number of employees per establishment.

Table 5.3a presents for each size code of the forging and stamping industry the number of establishments in that size code, the number of employees working at those establishments, and the average number of employees for the establishments in that size code, for the years 1990, 1995, 2000, 2005, 2008, 2010, and 2012. Note in Table 5.3a that there are no forging facilities with size code 9, 1,000 or more employees. For size code 8, 500 to 999 employees, the number of establishments grew from 10–20 between 1990 and 2000 but dropped from 20–7 by 2008 and then to 3 in 2010. The number then returned to the 2008 level of 7 as workers were recalled. For the next largest size code, 7, having between 250 and 499 employees, the number of establishments grew from 45 in 1990 to 74 in 2000 but then dropped to 55 in 2008 then to 28 in 2010. The number then increased to 45 in 2012 with the recovery. The next two size codes did not experience quite such large drops. The loss of facilities in the forging and stamping industry occurred more or less across all size codes.

Table 5.3b presents the change in employment at each establishment size code for the forging and stamping industry. As may be seen, this industry incurred losses of establishments at every level of size code and the employment losses were evenly distributed across size codes.

Table 5.4a presents for each size code of the metalworking machinery manufacturing industry the number of establishments in that size code, the number of employees working at those establishments, the average number of employees for the establishments in that size code, and the percentage of all establishments that size code represents for the years 1990, 1995, 2000, 2005, 2008, 2010, and 2012. Note in Table 5.4a that there are no facilities with size code 9, 1,000 or more employees. For size code 8, 500 to 999 employees, the number of establishments dropped from 25 to only 6 between 1990 and 2005. The number dropped to 4 in 2010, and then, following the recession, the number of size code 8 plants

Table 5.2b Foundries industry change in employment 1990–2012

Size Code	Change in Employment Due to Change in Establishments			Size Code	Change in Employment Due to Change in Emp/Est			Size Code	Change in Employment Due to Combined Effect			Size Code	Change in Total Employment
	(Δ Est)	(EPE.)	=		(Δ EPE)	(Est.)	=		(Δ Est)	(Δ EPE)	=		
1	-38	-2.3	-86	1	-0.5	532	-276	1	-38	-0.5	20	1	-342
2	-85	6.9	-588	2	-0.1	354	-50	2	-85	-0.1	12	2	-625
3	-142	14.1	-1,995	3	-0.1	446	-56	3	-142	-0.1	18	3	-2,034
4	-297	31.0	-9,219	4	1.4	686	985	4	-297	1.4	-426	4	-8,661
5	-122	71.0	-8,660	5	0.0	379	-18	5	-122	0.0	6	5	-8,672
6	-153	156.2	-23,903	6	-0.8	357	-272	6	-153	-0.8	117	6	-24,059
7	-21	339.2	-7,123	7	6.2	108	668	7	-21	6.2	-130	7	-6,585
8	-18	673.6	-12,125	8	20.5	47	962	8	-18	20.5	-368	8	-11,532
9	-16	1,593.2	-25,491	9	-80.7	21	-1,694	9	-16	-80.7	1,291	9	-25,895

TOTAL LOSS OF EMPLOYMENT   -88,404

Table 5.3a Distribution of establishment size for forging and stamping industry

Year	Size Code 1 < 5 # of Estab.	Employees	Emp/Est	Size Code 2 5–9 # of Estab.	Employees	Emp/Est	Size Code 3 10–19 # of Estab.	Employees	Emp/Est
1990	NA	NA	NA	447	3,033	6.8	502	7,031	14.0
1995	NA	NA	NA	416	2,840	6.8	444	6,219	14.0
2000	NA	NA	NA	397	2,716	6.8	412	5,834	14.2
2005	623*	NA	NA	353	2,379	6.7	436	6,059	13.9
2008	604*	NA	NA	375	2,568	6.8	410	5,802	14.2
2010	674*	NA	NA	368	2,484	6.8	424	5,843	13.8
2012	597*	NA	NA	342	2,321	6.8	408	5,762	14.1

Year	Size Code 4 20–49 # of Estab.	Employees	Emp/Est	Size Code 5 50–99 # of Estab.	Employees	Emp/Est	Size Code 6 100–249 # of Estab.	Employees	Emp/Est
1990	632	19,689	31.2	330	23,319	70.7	210	31,975	152.3
1995	604	18,844	31.2	355	24,778	69.8	231	36,292	157.1
2000	606	19,169	31.6	362	25,171	69.5	282	42,623	151.1
2005	586	18,359	31.3	343	24,012	70.0	209	31,288	149.7
2008	578	18,266	31.6	353	24,532	69.5	211	31,096	147.4

(Continued)

Table 5.3a (Continued)

Year	Size Code 4 20–49			Size Code 5 50–99			Size Code 6 100–249		
	# of Estab.	Employees	Emp/Est	# of Estab.	Employees	Emp/Est	# of Estab.	Employees	Emp/Est
2010	544	17,060	31.4	317	21,789	68.7	155	23,583	152.2
2012	537	17,104	31.9	308	21,031	68.3	195	29,099	149.2

Year	Size Code 7 250–499			Size Code 8 500–999			Size Code 9 1000 or more		
	# of Estab.	Employees	Emp/Est	# of Estab.	Employees	Emp/Est	# of Estab.	Employees	Emp/Est
1990	45	15,710	349.1	10	6,983	698.3	NA	NA	NA
1995	64	21,542	336.6	12	8,545	712.1	NA	NA	NA
2000	74	23,377	315.9	20	14,062	703.1	NA	NA	NA
2005	52	18,105	348.2	9	5,756	639.5	2*	NA	NA
2008	55	17,980	326.9	7	4,733	676.1	2*	NA	NA
2010	28	9,369	334.6	3	1,847	615.6	2*	NA	NA
2012	45	14,693	326.5	7	4,781	683.0	1*	NA	NA

Source: Bureau of Labor Statistics (2013).

1. * Establishment totals for time period had a quarterly status code of "N." Therefore, the Bureau of Labor Statistics includes establishment numbers in the report totals but employment figures are not and subsequently marked as "0" in the employment totals. Information regarding status/disclosure codes is located at the BLS website frp://ftp.bls.gov/pub/special.requests/cew/DOCUMENT/layout.txt.

2. An "NA" appearing in the number of establishments column for a specific size code denotes that the size code did not appear on the BLS report.

Table 5.3b Forging and stamping industry change in employment 1990–2012

Size Code	Change in Employment Due to Change in Establishments			Size Code	Change in Employment Due to Change in Emp/Est			Size Code	Change in Employment Due to Combined Effect			Size Code	Change in Total Employment
	(Δ Est)	(EPE.)	=		(Δ EPE)	(Est.)	=		(Δ Est)	(Δ EPE)	=		
1	NA	NA	NA	1	NA	NA	NA	1	NA	NA	NA	1	NA
2	−105	6.8	−712	2	0.0	447	1	2	−105	0.0	0	2	−712
3	−94	14.0	−1,316	3	0.1	502	59	3	−94	0.1	−11	3	−1,268
4	−95	31.2	−2,960	4	0.7	632	441	4	−95	0.7	−66	4	−2,585
5	−22	70.7	−1,555	5	−2.4	330	−785	5	−22	−2.4	52	5	−2,288
6	−15	152.3	−2,284	6	−3.0	210	−638	6	−15	−3.0	46	6	−2,876
7	0	349.1	0	7	−22.6	45	−1,017	7	0	−22.6	0	7	−1,017
8	−3	698.3	−2,095	8	−15.2	10	−152	8	−3	−15.2	46	8	−2,201
9	NA	NA	NA	9	NA	NA	NA	9	NA	NA	NA	9	NA

TOTAL LOSS OF EMPLOYMENT   −12,947

Table 5.4a Distribution of establishment size for metalworking machinery industry

| | Size Code 1 | | | Size Code 2 | | | Size Code 3 | | |
| | <5 | | | 5–9 | | | 10–19 | | |
Year	# of Estab.	Employees	Emp/Est	# of Estab.	Employees	Emp/Est	# of Estab.	Employees	Emp/Est
1990	4,087	8,355	2.0	2,726	18,470	6.8	2,591	35,563	13.7
1995	4,058	8,143	2.0	2,619	17,648	6.7	2,604	35,489	13.6
2000	4,658	8,848	1.9	2,614	17,502	6.7	2,505	34,121	13.6
2005	4,086	7,734	1.9	2,220	14,970	6.7	2,051	28,042	13.7
2008	3,673	7,046	1.9	1,996	13,500	6.8	1,873	25,548	13.6
2010	3,888	6,970	1.8	1,812	12,122	6.7	1,644	22,540	13.7
2012	3,274	6,239	1.9	1,763	11,771	6.7	1,597	21,795	13.6

| | Size Code 4 | | | Size Code 5 | | | Size Code 6 | | |
| | 20–49 | | | 50–99 | | | 100–249 | | |
Year	# of Estab.	Employees	Emp/Est	# of Estab.	Employees	Emp/Est	# of Estab.	Employees	Emp/Est
1990	1,920	57,858	30.1	638	43,931	68.9	327	47,699	145.9
1995	2,127	63,885	30.0	717	49,272	68.7	359	52,764	147.0
2000	2,178	65,544	30.1	813	55,859	68.7	367	53,360	145.4

Year	# of Estab.	Employees	Emp/Est	# of Estab.	Employees	Emp/Est	# of Estab.	Employees	Emp/Est
2005	1,658	49,924	30.1	607	41,083	67.7	260	37,699	145.0
2008	1,590	48,415	30.4	548	37,491	68.4	257	37,434	145.7
2010	1,300	39,091	30.1	395	26,592	67.3	201	28,979	144.2
2012	1,412	42,174	29.9	496	33,537	67.6	264	37,867	143.4

	Size Code 7 250–499			Size Code 8 500–999			Size Code 9 1000 or more		
Year	# of Estab.	Employees	Emp/Est	# of Estab.	Employees	Emp/Est	# of Estab.	Employees	Emp/Est
1990	71	24,439	344.2	25	17,780	711.2	NA	NA	NA
1995	61	20,592	337.6	25	16,644	665.8	NA	NA	NA
2000	65	22,201	341.5	24	15,405	641.9	NA	NA	NA
2005	51	16,972	332.8	6	3,718	619.7	NA	NA	NA
2008	47	15,755	335.2	10*	NA	NA	1*	NA	NA
2010	35	11,744	335.5	4	2,522	630.5	NA	NA	NA
2012	47	15,387	327.4	8	5,138	642.3	NA	NA	NA

Source: Bureau of Labor Statistics (2013).

1. * Establishment totals for time period had a quarterly status code of "N." Therefore, the Bureau of Labor Statistics includes establishment numbers in the report totals but employment figures are not and subsequently marked as "0" in the employment totals. Information regarding status/disclosure codes is located at the BLS website ftp://ftp.bls.gov/pub/special.requests/cew/DOCUMENT/layout.txt.

2. An "NA" appearing in the number of establishments column for a specific size code denotes that the size code did not appear on the BLS report.

increased to 8. For the next largest size code, 7, having between 250 and 499 employees, the number of establishments dropped from 71 in 1990 to 35 in 2010 but rebounded to 47 after the recession. The next two size codes did not experience quite such large drops. For this industry, there was a loss of employment and of establishments at every size code as may be seen in Table 5.4b. Loss of employment was in the same range for size codes 3 to 8, a somewhat different pattern than for other durable industries.

Table 5.5a presents for each size code of the household appliance manufacturing industry the number of establishments in that size code, the number of employees working at those establishments, and the average number of employees for the establishments in that size code, for the years 1990, 1995, 2000, 2005, 2008, 2010, and 2012. Note in Table 5.5a that for size code 9, with 1,000 or more employees, the number of establishments dropped from 28 to 16 between 1990 and 2010. For this industry there was no rebound following the recession but a further drop to 15. For the next largest size code, 8, having between 500 and 999 employees, the number of establishments dropped from 40 to 14 in 2010 and then a subsequent drop to 12 in 2012. The next size code, 7, also experienced large drops in the number of establishments, dropping from 37 to 17 in 2010 and increasing to 22 in 2012. As may be seen in Table 5.5b, the largest loss of jobs occurred by far in the two largest size codes.

Table 5.6a presents for each size code of the computer and peripheral equipment manufacturing industry the number of establishments in that size code, the number of employees working at those establishments, and the average number of employees for the establishments in that size code, for the years 1990, 1995, 2000, 2005, 2008, 2010, and 2012. Note in Table 5.6a that for size code 9, 1,000 or more employees, the number of establishments dropped from 66 to only 32 between 1990 and 2010, and further dropped to 28 during the recovery. This closing of establishments resulted in a loss of 102,000 jobs. For the next largest size code, 8, having between 500 and 999 employees, the number of establishments experienced a precipitous decline, dropping from 71 to 23 in 2010 with a growth to 24 in 2012. These closings resulted in the loss of another 35,000 jobs. The next two size codes did not experience quite such large

Table 5.4b Metalworking machinery industry change in employment 1990–2012

	Change in Employment Due to Change in Establishments				Change in Employment Due to Change in Emp/Est				Change in Employment Due to Combined Effect				Change in Total Employment
Size Code	(Δ Est)	(EPE.)	=	Size Code	(Δ EPE)	(Est.)	=	Size Code	(Δ Est)	(Δ EPE)	=	Size Code	
1	-813	-2.0	-1,662	1	-0.1	4,087	-566	1	-813	-0.1	113	1	-2,115
2	-963	6.8	-6,525	2	-0.1	2,726	-270	2	-963	-0.1	95	2	-6,699
3	-994	13.7	-13,643	3	-0.1	2,591	-203	3	-994	-0.1	78	3	-13,768
4	-508	30.1	-15,308	4	-0.3	1,920	-511	4	-508	-0.3	135	4	-15,684
5	-142	68.9	-9,778	5	-1.2	638	-793	5	-142	-1.2	176	5	-10,394
6	-63	145.9	-9,190	6	-2.4	327	-795	6	-63	-2.4	153	6	-9,832
7	-24	344.2	-8,261	7	-16.8	71	-1,194	7	-24	-16.8	404	7	-9,052
8	-17	711.2	-12,090	8	-68.9	25	-1,723	8	-17	-68.9	1,171	8	-12,642
9	NA	NA	NA	9	NA	NA	NA	9	NA	NA	NA	9	NA

TOTAL LOSS OF EMPLOYMENT   -80,186

Table 5.5a Distribution of establishment size for household appliances industry

Year	Size Code 1 <5			Size Code 2 5–9			Size Code 3 10–19		
	# of Estab.	Employees	Emp/Est	# of Estab.	Employees	Emp/Est	# of Estab.	Employees	Emp/Est
1990	124	218	1.8	45	304	6.8	50	688	13.8
1995	152	253	1.7	53	369	7.0	47	636	13.5
2000	196	496	2.5	62	412	6.6	60	801	13.4
2005	194	328	1.7	90	637	7.1	49	642	13.1
2008	194	306	1.6	68	459	6.8	57	781	13.7
2010	186	281	1.5	65	446	6.9	48	649	13.5
2012	187	305	1.6	71	480	6.8	47	667	14.2

Year	Size Code 4 20–49			Size Code 5 50–99			Size Code 6 100–249		
	# of Estab.	Employees	Emp/Est	# of Estab.	Employees	Emp/Est	# of Estab.	Employees	Emp/Est
1990	46	1,470	32.0	22	1,521	69.1	53	9,711	183.2
1995	50	1,838	36.8	35	2,235	63.9	49	8,562	174.7

2000	54	1,747	32.4	41	3,006	73.3	40	6,418	160.5
2005	64	2,114	33.0	34	2,479	72.9	45	7,692	170.9
2008	59	1,934	32.8	33	2,321	70.3	45	7,081	157.3
2010	54	1,686	31.2	30	2,013	67.1	44	6,920	157.3
2012	45	1,368	30.4	31	2,174	70.1	37	5,771	156.0

Year	Size Code 7 250–499			Size Code 8 500–999			Size Code 9 1000 or more		
	# of Estab.	Employees	Emp/Est	# of Estab.	Employees	Emp/Est	# of Estab.	Employees	Emp/Est
1990	37	12,725	343.9	40	29,583	739.6	28	55,331	1,976.1
1995	34	12,385	364.3	30	21,710	723.7	31	61,139	1,972.2
2000	39	13,795	353.7	26	17,331	666.6	32	65,781	2,055.7
2005	30	10,995	366.5	19	13,046	686.6	26	49,603	1,907.8
2008	26	8,924	343.2	13	9,683	744.8	22	40,464	1,839.3
2010	17	5,538	325.7	14	10,860	775.7	16	29,163	1,822.7
2012	22	7,160	325.5	12	9,240	770.0	15	28,106	1,873.8

Source: Bureau of Labor Statistics (2013).

Table 5.5b Household appliances industry change in employment 1990–2012

Size Code	Change in Employment Due to Change in Establishments			Size Code	Change in Employment Due to Change in Emp/Est			Size Code	Change in Employment Due to Combined Effect			Size Code	Change in Total Employment
	(Δ Est)	(EPE.)	=		(Δ EPE)	(Est.)	=		(Δ Est)	(Δ EPE)	=		
1	63	-1.8	111	1	-0.1	124	-16	1	63	-0.1	-8	1	87
2	26	6.8	176	2	0.0	45	0	2	26	0.0	0	2	176
3	-3	13.8	-41	3	0.4	50	22	3	-3	0.4	-1	3	-21
4	-1	32.0	-32	4	-1.6	46	-72	4	-1	-1.6	2	4	-103
5	9	69.1	622	5	1.0	22	22	5	9	1.0	9	5	653
6	-16	183.2	-2,932	6	-27.3	53	-1,444	6	-16	-27.3	436	6	-3,940
7	-15	343.9	-5,159	7	-18.4	37	-682	7	-15	-18.4	277	7	-5,564
8	-28	739.6	-20,708	8	30.4	40	1,216	8	-28	30.4	-851	8	-20,343
9	-13	1,976.1	-25,689	9	-102.4	28	-2,866	9	-13	-102.4	1,331	9	-27,225

TOTAL LOSS OF EMPLOYMENT   -56,280

Table 5.6a Distribution of establishment size for computer and peripheral equipment industry

| | Size Code 1 | | | Size Code 2 | | | Size Code 3 | | |
| | < 5 | | | 5–9 | | | 10–19 | | |
Year	# of Estab.	Employees	Emp/Est	# of Estab.	Employees	Emp/Est	# of Estab.	Employees	Emp/Est
1990	585	1,150	2.0	244	1,661	6.8	246	3,430	13.9
1995	835	1,373	1.6	254	1,753	6.9	267	3,645	13.7
2000	850	1,273	1.5	290	1,938	6.7	261	3,628	13.9
2005	686	1,061	1.5	215	1,437	6.7	212	2,986	14.1
2008	691	1,142	1.7	202	1,360	6.7	194	2,703	13.9
2010	731	1,054	1.4	185	1,214	6.6	183	2,489	13.6
2012	742	1,070	1.4	181	1,195	6.6	183	2,500	13.7

| | Size Code 4 | | | Size Code 5 | | | Size Code 6 | | |
| | 20–49 | | | 50–99 | | | 100–249 | | |
Year	# of Estab.	Employees	Emp/Est	# of Estab.	Employees	Emp/Est	# of Estab.	Employees	Emp/Est
1990	277	8,906	32.2	181	12,562	69.4	175	27,462	156.9
1995	330	10,487	31.8	161	11,312	70.3	163	25,247	154.9
2000	328	10,341	31.5	212	14,582	68.8	164	25,605	156.1

(Continued)

Table 5.6a (Continued)

Year	Size Code 4 20-49			Size Code 5 50-99			Size Code 6 100-249		
	# of Estab.	Employees	Emp/Est	# of Estab.	Employees	Emp/Est	# of Estab.	Employees	Emp/Est
2005	283	9,108	32.2	124	8,738	70.5	121	19,436	160.6
2008	226	6,949	30.7	123	8,442	68.6	123	19,065	155.0
2010	215	6,845	31.8	101	6,990	69.2	101	15,618	154.6
2012	189	6,087	32.2	101	7,203	71.3	97	16,060	165.6

Year	Size Code 7 250-499			Size Code 8 500-999			Size Code 9 1000 or more		
	# of Estab.	Employees	Emp/Est	# of Estab.	Employees	Emp/Est	# of Estab.	Employees	Emp/Est
1990	74	27,355	369.7	71	51,920	731.3	66	188,016	2,848.7
1995	73	24,591	336.9	56	40,289	719.4	53	138,138	2,606.4
2000	91	31,877	350.3	58	41,113	708.9	58	156,793	2,703.3
2005	58	20,040	345.5	42	29,278	697.1	40	110,869	2,771.7
2008	62	22,208	358.2	21	14,204	676.4	39	106,684	2,735.5
2010	53	17,941	338.5	23	15,646	680.3	32	91,042	2,845.1
2012	57	20,126	353.1	24	16,775	698.9	28	85,767	3,063.1

Source: Bureau of Labor Statistics (2013).

drops. As may be seen in Table 5.6b, the loss of facilities in the computer and peripheral equipment manufacturing industry again occurred primarily at larger facilities.

Table 5.7a presents for each size code of the communications equipment manufacturing industry the number of establishments in that size code, the number of employees working at those establishments, and the average number of employees for the establishments in that size code, for the years 1990, 1995, 2000, 2005, 2008, 2010, and 2012. Note in Table 5.7a that for size code 9, 1,000 or more employees, the number of establishments dropped from 53 to only 16 between 1990 and 2010 and then dropped to 15 in 2012. These plant closings resulted in the loss of 106,000 jobs. For the next largest size code, 8, having between 500 and 999 employees, the number of establishments experienced a significant decline, dropping from 49 to 27 in 2010 and then a drop to 23 in 2012. Again, it is seen that it is the largest plants that were closed. The next two size codes did not experience quite such large drops. The loss of facilities and communications equipment manufacturing plants again occurred primarily at larger facilities as shown in Table 5.7b. Two thirds of the total job losses occurred at the largest size code.

Table 5.8a presents for each size code of the semiconductor and electronic components manufacturing industry the number of establishments in that size code, the number of employees working at those establishments, and the average number of employees for the establishments in that size code, for the years 1990, 1995, 2000, 2005, 2008, and 2010. Note in Table 5.8a that for size code 9, 1,000 or more employees, the number of establishments dropped from 72 to 42 between 1990 and 2010, and then recovered slightly to 46 establishments in 2012 as workers were recalled. Nevertheless, over the time period from 1990 to 2012, the closing of size code 9 establishments resulted in the loss of 72,000 jobs. For the next largest size code, 8, having between 500 and 999 employees, the number of establishments experienced a significant decline, dropping from 113 to 66 between 1990 and 2010, and then dropped to 63 in 2012. Overall, this size code lost 37,000 employees. The next two size codes did not experience quite such large drops but experienced significant job losses of 26,000

Table 5.6b Computer and peripheral equipment industry change in employment 1990–2012

Size Code	Change in Employment Due to Change in Establishments			Size Code	Change in Employment Due to Change in Emp/Est			Size Code	Change in Employment Due to Combined Effect			Size Code	Change in Total Employment
	(Δ Est)	(EPE.)	=		(Δ EPE)	(Est.)	=		(Δ Est)	(Δ EPE)	=		
1	157	-2.0	309	1	-0.5	585	-306	1	157	-0.5	-82	1	-79
2	-63	6.8	-429	2	-0.2	244	-50	2	-63	-0.2	13	2	-466
3	-63	13.9	-879	3	-0.3	246	-69	3	-63	-0.3	18	3	-930
4	-88	32.2	-2,829	4	0.1	277	15	4	-88	0.1	-5	4	-2,819
5	-80	69.4	-5,552	5	1.9	181	345	5	-80	1.9	-153	5	-5,360
6	-78	156.9	-12,240	6	8.6	175	1,512	6	-78	8.6	-674	6	-11,402
7	-17	369.7	-6,284	7	-16.6	74	-1,226	7	-17	-16.6	282	7	-7,229
8	-47	731.3	-34,369	8	-32.3	71	-2,295	8	-47	-32.3	1,519	8	-35,145
9	-38	2,848.7	-108,252	9	214.4	66	14,149	9	-38	214.4	-8,146	9	-102,249

TOTAL LOSS OF EMPLOYMENT   -165,679

Table 5.7a Distribution of establishment size for communication equipment industry

| | Size Code 1 | | | Size Code 2 | | | Size Code 3 | | |
| | < 5 | | | 5–9 | | | 10–19 | | |
Year	# of Estab.	Employees	Emp/Est	# of Estab.	Employees	Emp/Est	# of Estab.	Employees	Emp/Est
1990	578	1,024	1.8	353	2,361	6.7	302	4,245	14.1
1995	824	1,353	1.6	357	2,353	6.6	376	5,185	13.8
2000	1,055	1,832	1.7	365	2,478	6.8	365	5,068	13.9
2005	887	1,652	1.9	345	2,323	6.7	302	4,222	14.0
2008	844	1,351	1.6	295	1,993	6.8	284	3,977	14.0
2010	859	1,254	1.5	268	1,827	6.8	276	3,907	14.2
2012	942	1,509	1.6	312	2,132	6.8	304	4,305	14.2

| | Size Code 4 | | | Size Code 5 | | | Size Code 6 | | |
| | 20–49 | | | 50–99 | | | 100–249 | | |
Year	# of Estab.	Employees	Emp/Est	# of Estab.	Employees	Emp/Est	# of Estab.	Employees	Emp/Est
1990	421	13,333	31.7	236	16,685	70.7	207	32,368	156.4
1995	435	13,708	31.5	258	18,003	69.8	225	34,783	154.6
2000	448	14,197	31.7	268	18,714	69.8	266	40,563	152.5

(Continued)

Table 5.7a (Continued)

Year	Size Code 4 20–49			Size Code 5 50–99			Size Code 6 100–249		
	# of Estab.	Employees	Emp/Est	# of Estab.	Employees	Emp/Est	# of Estab.	Employees	Emp/Est
2005	379	12,009	31.7	223	15,723	70.5	197	30,499	154.8
2008	325	10,206	31.4	186	13,151	70.7	172	26,831	156.0
2010	323	10,146	31.4	168	11,918	70.9	156	24,263	155.5
2012	314	9,872	31.4	172	12,401	72.1	143	22,510	157.4

Year	Size Code 7 250–499			Size Code 8 500–999			Size Code 9 1000 or more		
	# of Estab.	Employees	Emp/Est	# of Estab.	Employees	Emp/Est	# of Estab.	Employees	Emp/Est
1990	81	27,335	337.5	49	32,858	670.6	53	133,001	2,509.5
1995	103	36,014	349.6	39	26,751	685.9	42	100,082	2,382.9
2000	110	37,794	343.6	50	34,162	683.2	44	85,734	1,948.5
2005	74	26,824	362.5	25	17,319	692.7	22	36,093	1,640.6
2008	61	20,558	337.0	28	19,972	713.3	17	29,363	1,727.2
2010	55	18,087	328.9	27	18,430	682.6	16	26,754	1,672.1
2012	48	16,426	342.2	23	14,988	651.7	15	26,611	1,774.0

Source: Bureau of Labor Statistics (2013).

Table 5.7b *Communications equipment industry change in employment 1990–2012*

Size Code	Change in Employment Due to Change in Establishments			Size Code	Change in Employment Due to Change in Emp/Est			Size Code	Change in Employment Due to Combined Effect			Size Code	Change in Total Employment
	(Δ Est)	(EPE.)	=		(Δ EPE)	(Est.)	=		(Δ Est)	(Δ EPE)	=		
1	364	-1.8	645	1	-0.2	578	-99	1	364	-0.2	-62	1	484
2	-41	6.7	-274	2	0.1	353	50	2	-41	0.1	-6	2	-230
3	2	14.1	28	3	0.1	302	32	3	2	0.1	0	3	60
4	-107	31.7	-3,389	4	-0.2	421	-97	4	-107	-0.2	25	4	-3,461
5	-64	70.7	-4,525	5	1.4	236	330	5	-64	1.4	-89	5	-4,284
6	-64	156.4	-10,008	6	1.0	207	216	6	-64	1.0	-67	6	-9,858
7	-33	337.5	-11,136	7	4.7	81	383	7	-33	4.7	-156	7	-10,909
8	-26	670.6	-17,435	8	-18.9	49	-927	8	-26	-18.9	492	8	-17,870
9	-38	2,509.5	-95,359	9	-735.4	53	-38,977	9	-38	-735.4	27,946	9	-106,391

TOTAL LOSS OF EMPLOYMENT   -152,460

Table 5.8a Distribution of establishment size for semiconductors and electronic components industry

| | Size Code 1 | | | Size Code 2 | | | Size Code 3 | | |
| | < 5 | | | 5–9 | | | 10–19 | | |
Year	# of Estab.	Employees	Emp/Est	# of Estab.	Employees	Emp/Est	# of Estab.	Employees	Emp/Est
1990	1,486	2,913	2.0	811	5,372	6.6	891	12,303	13.8
1995	1,688	3,013	1.8	845	5,616	6.6	997	13,722	13.8
2000	1,917	3,257	1.7	814	5,540	6.8	946	13,121	13.9
2005	1,721	2,911	1.7	788	5,861	7.4	926	13,016	14.1
2008	1,613	3,152	2.0	786	5,390	6.9	807	11,243	13.9
2010	1,727	3,069	1.8	763	5,270	6.9	825	11,457	13.9
2012	1,686	3,086	1.8	755	5,251	7.0	818	11,410	13.9

| | Size Code 4 | | | Size Code 5 | | | Size Code 6 | | |
| | 20–49 | | | 50–99 | | | 100–249 | | |
Year	# of Estab.	Employees	Emp/Est	# of Estab.	Employees	Emp/Est	# of Estab.	Employees	Emp/Est
1990	1,209	38,142	31.5	776	54,886	70.7	669	102,937	153.9
1995	1,276	39,892	31.3	865	60,969	70.5	627	96,635	154.1

2000	1,313	41,373	31.5	855	60,069	70.3	701	106,481	151.9
2005	1,159	36,666	31.6	727	51,877	71.4	536	83,548	155.9
2008	1,141	36,166	31.7	695	49,339	71.0	549	85,570	155.9
2010	1,096	34,988	31.9	604	42,489	70.3	463	68,918	148.9
2012	1,057	34,067	32.2	605	42,583	70.4	529	80,187	151.6

Year	Size Code 7 250-499			Size Code 8 500-999			Size Code 9 1000 or more		
	# of Estab.	Employees	Emp/Est	# of Estab.	Employees	Emp/Est	# of Estab.	Employees	Emp/Est
1990	219	77,456	353.7	113	78,779	697.2	72	187,648	2,606.2
1995	225	76,964	342.1	117	80,197	685.4	65	167,960	2,584.0
2000	263	90,305	343.4	133	91,126	685.2	92	235,004	2,554.4
2005	177	60,690	342.9	88	60,255	684.7	58	133,978	2,310.0
2008	172	58,464	339.9	85	58,266	685.5	55	130,017	2,363.9
2010	163	55,571	340.9	66	44,562	675.2	42	98,088	2,335.4
2012	151	51,230	339.3	63	41,527	659.2	46	115,474	2,510.3

Source: Bureau of Labor Statistics (2013).

and 23,000. The loss of jobs again occurred primarily at larger facilities as seen in Table 5.8b.

Table 5.9a presents for each size code of the aerospace products and parts manufacturing industry the number of establishments in that size code, the number of employees working at those establishments, and the average number of employees for the establishments in that size code, for the years 1990, 1995, 2000, 2005, 2008, 2010, and 2012. This industry has a very different pattern than seen so far for other durable goods industries. It is important to note that the total employment in the largest size code dropped from 662,074 to just 334,135 between 1990 and the year 2000, a loss of 327,939 jobs. Also during the decade of the 90s, 37 large (greater than 1,000 employees) establishments were closed, each averaging 5,000 to 6,000 employees (see Articles 5.1, 5.2, and 5.3). This enormous decline can be traced to the large reductions in defense spending following the end of the Cold War in what has been referred to as the "peace dividend." This decline in defense expenditures may be seen in Figure 5.1. The next smaller size code groups experienced modest gains in the number of establishments and employment during the period between 1990 and 2000. The strong growth in defense expenditures following 9/11 led to growth in the number of establishments between 2000 and 2008. Nevertheless, between 1990 and 2012, for the aerospace products and parts manufacturing industry, the loss of facilities and jobs occurred dramatically at the largest facilities during the 90s. The loss by size code is presented in Table 5.9b.

Table 5.10a presents for each size code of the household and institutional furniture manufacturing industry the number of establishments in that size code, the number of employees working at those establishments, and the average number of employees for the establishments in that size code, for the years 1990, 1995, 2000, 2005, 2008, 2010, and 2012. This industry has been hit hard at every size code. In Table 5.10a for size code 9, 1,000 or more employees, the number of establishments dropped from 28 to only 9 between 1990 and 2010, and then dropped to 8 in 2012. For the next largest size code, 8, having between 500 and 999 employees, the number of establishments experienced a significant decline, dropping from 71 to 36. For household and institutional furniture manufacturers,

Table 5.8b Semiconductors and electronic components industry change in employment 1990–2012

Size Code	Change in Employment Due to Change in Establishments			Size Code	Change in Employment Due to Change in Emp/Est			Size Code	Change in Employment Due to Combined Effect			Size Code	Change in Total Employment
	(Δ Est)	(EPE.)	=		(Δ EPE)	(Est.)	=		(Δ Est)	(Δ EPE)	=		
1	200	-2.0	392	1	-0.1	1,486	-193	1	200	-0.1	-26	1	173
2	-56	6.6	-371	2	0.3	811	268	2	-56	0.3	-19	2	-121
3	-73	13.8	-1,008	3	0.1	891	125	3	-73	0.1	-10	3	-893
4	-152	31.5	-4,795	4	0.7	1,209	823	4	-152	0.7	-103	4	-4,076
5	-171	70.7	-12,095	5	-0.3	776	-267	5	-171	-0.3	59	5	-12,303
6	-140	153.9	-21,541	6	-2.3	669	-1,529	6	-140	-2.3	320	6	-22,750
7	-68	353.7	-24,050	7	-14.4	219	-3,155	7	-68	-14.4	980	7	-26,225
8	-50	697.2	-34,858	8	-38.0	113	-4,294	8	-50	-38.0	1,900	8	-37,252
9	-26	2,606.2	-67,762	9	-95.9	72	-6,906	9	-26	-95.9	2,494	9	-72,174

TOTAL LOSS OF EMPLOYMENT   -175,621

Table 5.9a Distribution of establishment size for aerospace products and parts industry

	Size Code 1 < 5			Size Code 2 5–9			Size Code 3 10–19		
Year	# of Estab.	Employees	Emp/Est	# of Estab.	Employees	Emp/Est	# of Estab.	Employees	Emp/Est
1990	506	1,035	2.0	324	2,153	6.6	425	5,820	13.7
1995	664	1,352	2.0	378	2,547	6.7	403	5,593	13.9
2000	760	1,400	1.8	358	2,494	7.0	412	5,615	13.6
2005	830	1,400	1.7	371	2,554	6.9	376	5,198	13.8
2008	953	2,019	2.1	355	2,409	6.8	369	5,277	14.3
2010	945	1,890	2.0	352	2,410	6.8	391	5,475	14.0
2012	927	1,609	1.7	388	2,538	6.5	380	5,373	14.1

	Size Code 4 20–49			Size Code 5 50–99			Size Code 6 100–249		
Year	# of Estab.	Employees	Emp/Est	# of Estab.	Employees	Emp/Est	# of Estab.	Employees	Emp/Est
1990	489	15,401	31.5	280	19,544	69.8	234	36,371	155.4
1995	515	15,838	30.8	269	18,833	70.0	255	40,056	157.1

	# of Estab.	Employees	Emp/Est	# of Estab.	Employees	Emp/Est	# of Estab.	Employees	Emp/Est
2000	506	16,065	31.7	286	20,393	71.3	292	45,122	154.5
2005	483	15,265	31.6	281	20,131	71.6	282	43,802	155.3
2008	464	14,988	32.3	278	19,945	71.7	309	47,802	154.7
2010	460	14,763	32.1	289	20,694	71.6	284	43,782	154.2
2012	493	15,786	32.0	277	19,617	70.8	304	45,438	149.5

Year	Size Code 7 250–499			Size Code 8 500–999			Size Code 9 1000 or more		
	# of Estab.	Employees	Emp/Est	# of Estab.	Employees	Emp/Est	# of Estab.	Employees	Emp/Est
1990	120	42,447	353.7	57	40,288	706.8	108	662,074	6,130.3
1995	97	33,924	349.7	63	44,343	703.9	74	357,846	4,835.8
2000	131	45,587	348.0	72	48,581	674.7	71	334,135	4,706.1
2005	106	36,681	346.1	70	47,575	679.6	68	273,375	4,020.2
2008	130	45,373	349.0	86	58,475	679.9	76	304,475	4,006.3
2010	129	46,295	358.9	79	53,367	675.5	70	287,192	4,102.7
2012	146	51,939	355.7	76	51,310	675.1	73	297,965	4,081.7

Source: Bureau of Labor Statistics (2013).

Table 5.9b *Aerospace products and parts industry change in employment 1990–2012*

Size Code	Change in Employment Due to Change in Establishments			Size Code	Change in Employment Due to Change in Emp/Est			Size Code	Change in Employment Due to Combined Effect			Size Code	Change in Total Employment
	(Δ Est)	(EPE.)	=		(Δ EPE)	(Est.)	=		(Δ Est)	(Δ EPE)	=		
1	421	-2.0	861	1	-0.3	506	-156	1	421	-0.3	-130	1	574
2	64	6.6	425	2	-0.1	324	-34	2	64	-0.1	-7	2	385
3	-45	13.7	-616	3	0.4	425	189	3	-45	0.4	-20	3	-448
4	4	31.5	126	4	0.5	489	257	4	4	0.5	2	4	385
5	-3	69.8	-209	5	1.0	280	285	5	-3	1.0	-3	5	73
6	70	155.4	10,880	6	-6.0	234	-1,395	6	70	-6.0	-417	6	9,068
7	26	353.7	9,197	7	2.0	120	243	7	26	2.0	53	7	9,492
8	19	706.8	13,429	8	-31.7	57	-1,806	8	19	-31.7	-602	8	11,022
9	-35	6,130.3	-214,561	9	-2,048.6	108	-221,249	9	-35	-2,048.6	71,701	9	-364,109

TOTAL LOSS OF EMPLOYMENT   -333,558

Article 5.1. Grumman; http://en.wikipedia.org/wiki/Grumman

At its peak in 1986 Grumman employed 23,000 people on Long Island and occupied 6,000,000 square feet (560,000 m²) in structures on 105 acres.

The end of the Cold War, at the beginning of the 1990s, the reduced need for defense spending led to a wave of mergers as aerospace companies shrank in number; in 1994 Northrop bought Grumman for $2.1 billion to form Northrop Grumman.

The new company closed almost all of its facilities on Long Island with the Bethpage plant being converted to a residential and office complex. Northrop Grumman's remaining business at the Bethpage campus is the Battle Management and Engagement Systems Division, which employs around 2,000 people.

Article 5.2. Closure of Hughes plant in Fullerton to be announced

Most of the complex's 6,800 workers will either be transferred or laid off, informed sources say; Don Lee and Ken Ellingwood, Times Staff Writers; Los Angeles Times, September 12, 1994

FULLERTON—Hughes Aircraft Co. will announce today that its sprawling Fullerton aerospace complex will be effectively shut down over the next two years, with most of the plant's 6,800 workers either transferred or laid off, according to sources knowledgeable about the plans.

"An awful lot of good people aren't going to survive this," said one well-placed source, adding that some workers would be transferred to Hughes sites in El Segundo and Long Beach while others will retire or lose their jobs. The source didn't know how many layoffs are anticipated at the Fullerton facility, which opened in 1957.

Article 5.3. Aerospace firm will close Santa Ana plant; SPS Technologies plans to lay off 300 workers by the end of 1993 because of sagging orders for aircraft parts; Ted Johnson; Los Angeles Times, Special to The Times; September 29, 1992

SANTA ANA—SPS Technologies Inc. said Monday that it will close its Santa Ana aerospace fastener plant by the end of 1993 and lay off 300 of the plant's 375 workers.

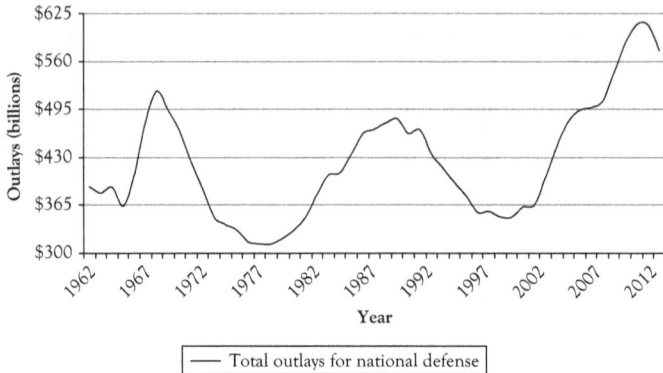

Figure 5.1 Total outlays for defense in constant 2005 dollars

the significant declines extended as well to the next three size codes. For size code 7, from 250 to 499 employees, the decline was from 211 to 725, a loss of 66 percent of the factories. For size code 6, from 100 to 249 employees, the number of establishments dropped from 514 to 281, a 46 percent reduction. For size code 5, from 50 to 99 employees, the number of establishments dropped from 710 to 392, a 44 percent reduction. Although the loss of plants again occurred primarily at larger facilities, the losses extended to the much smaller facilities for this industry. As may be seen in Table 5.10b, the largest loss of jobs occurred in size code 7, plants with 250 to 299 employees.

Table 5.11a presents for each size code of the office furniture and fixtures manufacturing industry the number of establishments in that size code, the number of employees working at those establishments, the average number of employees for the establishments in that size code, and the percentage of all establishments that size code represents for the years 1990, 1995, 2000, 2005, 2008, 2010, and 2012. Note in Table 5.11a that for size code 9, 1,000 or more employees, the number of establishments dropped from 10 to 5, between 1990 and 2012. For the next largest size code, 8, having between 500 and 999 employees, the number of establishments experienced a significant decline, dropping from 25 to 7. For office furniture and fixtures manufacturers, the significant declines extended as well to the next two size codes. For size code 7, from 250 to 499 employees, the decline was from 78 to 35. For size code 6, from

Table 5.10a Distribution of establishment size for household and institutional furniture industry

	Size Code 1 < 5			Size Code 2 5–9			Size Code 3 10–19		
Year	# of Estab.	Employees	Emp/Est	# of Estab.	Employees	Emp/Est	# of Estab.	Employees	Emp/Est
1990	10,328	19,902	1.9	4,209	27,423	6.5	2,590	34,305	13.2
1995	10,805	20,340	1.9	4,012	25,964	6.5	2,504	33,088	13.2
2000	10,360	18,579	1.8	3,934	25,541	6.5	2,782	37,235	13.4
2005	9,294	17,096	1.8	3,700	24,311	6.6	2,614	34,648	13.3
2008	9,177	17,396	1.9	3,667	24,338	6.6	2,391	32,266	13.5
2010	9,229	14,860	1.6	2,838	18,662	6.6	1,743	23,228	13.3
2012	7,787	13,014	1.7	2,498	16,160	6.5	1,605	21,223	13.2

	Size Code 4 20–49			Size Code 5 50–99			Size Code 6 100–249		
Year	# of Estab.	Employees	Emp/Est	# of Estab.	Employees	Emp/Est	# of Estab.	Employees	Emp/Est
1990	1,719	51,641	30.0	710	48,983	69.0	514	78,528	152.8
1995	1,634	49,351	30.2	636	44,074	69.3	472	74,477	157.8
2000	1,729	52,605	30.4	672	46,982	69.9	497	76,919	154.8

(Continued)

Table 5.10a (Continued)

Year	Size Code 4 20–49			Size Code 5 50–99			Size Code 6 100–249		
	# of Estab.	Employees	Emp/Est	# of Estab.	Employees	Emp/Est	# of Estab.	Employees	Emp/Est
2005	1,666	49,826	29.9	596	41,485	69.6	404	61,688	152.7
2008	1,565	47,337	30.2	523	36,618	70.0	355	54,736	154.2
2010	1,066	32,025	30.0	403	27,479	68.2	280	41,699	148.9
2012	1,029	30,414	29.6	392	27,387	69.9	281	41,595	148.0

Year	Size Code 7 250–499			Size Code 8 500–999			Size Code 9 1000 or more		
	# of Estab.	Employees	Emp/Est	# of Estab.	Employees	Emp/Est	# of Estab.	Employees	Emp/Est
1990	211	73,683	349.2	71	47,130	663.8	28	42,753	1,526.9
1995	206	71,624	347.7	80	52,087	651.1	25	36,045	1,441.8
2000	226	77,932	344.8	93	63,104	678.5	27	45,484	1,684.6
2005	172	58,765	341.7	74	48,478	655.1	27	43,073	1,595.3
2008	133	46,485	349.5	55	37,947	689.9	17	27,109	1,594.6
2010	72	24,672	342.7	40	26,102	652.6	9	14,760	1,640.0
2012	75	26,395	351.9	36	24,532	681.5	8	13,098	1,637.3

Source: Bureau of Labor Statistics (2013).

MANUFACTURING PLANT CLOSINGS BY SIZE   105

Table 5.10b Household and institutional furniture industry change in employment 1990–2012

Size Code	Change in Employment Due to Change in Establishments			Size Code	Change in Employment Due to Change in Emp/Est			Size Code	Change in Employment Due to Combined Effect			Size Code	Change in Total Employment
	(Δ Est)	(EPE.)	=		(Δ EPE)	(Est.)	=		(Δ Est)	(Δ EPE)	=		
1	-2,541	-1.9	-4,897	1	-0.3	10,328	-2,642	1	-2,541	-0.3	650	1	-6,888
2	-1,711	6.5	-11,148	2	0.0	4,209	-194	2	-1,711	0.0	79	2	-11,263
3	-985	13.2	-13,047	3	0.0	2,590	-57	3	-985	0.0	22	3	-13,082
4	-690	30.0	-20,728	4	-0.5	1,719	-833	4	-690	-0.5	334	4	-21,227
5	-318	69.0	-21,939	5	0.9	710	622	5	-318	0.9	-279	5	-21,595
6	-233	152.8	-35,597	6	-4.8	514	-2,443	6	-233	-4.8	1,107	6	-36,933
7	-136	349.2	-47,492	7	2.7	211	574	7	-136	2.7	-370	7	-47,288
8	-35	663.8	-23,233	8	17.7	71	1,254	8	-35	17.7	-618	8	-22,597
9	-20	1,526.9	-30,538	9	110.4	28	3,091	9	-20	110.4	-2,208	9	-29,655

TOTAL LOSS OF EMPLOYMENT   -210,529

Table 5.11a Distribution of establishment size for office furniture and fixtures industry

	Size Code 1 <5			Size Code 2 5–9			Size Code 3 10–19		
Year	# of Estab.	Employees	Emp/Est	# of Estab.	Employees	Emp/Est	# of Estab.	Employees	Emp/Est
1990	1,206	2,542	2.1	756	5,092	6.7	834	11,468	13.8
1995	1,354	2,697	2.0	759	5,077	6.7	792	10,943	13.8
2000	1,345	2,391	1.8	750	4,995	6.7	761	10,647	14.0
2005	1,498	2,705	1.8	715	4,805	6.7	680	9,319	13.7
2008	1,461	2,807	1.9	714	4,836	6.8	711	9,796	13.8
2010	1,583*	NA	NA	688	4,609	6.7	637	8,872	13.9
2012	1,440	2,466	1.7	608	4,060	6.7	591	8,225	13.9

	Size Code 4 20–49			Size Code 5 50–99			Size Code 6 100–249		
Year	# of Estab.	Employees	Emp/Est	# of Estab.	Employees	Emp/Est	# of Estab.	Employees	Emp/Est
1990	834	26,109	31.3	374	25,926	69.3	285	43,350	152.1
1995	845	25,795	30.5	406	28,607	70.5	246	37,297	151.6
2000	799	24,955	31.2	368	25,463	69.2	273	41,070	150.4

Year									
2005	727	22,271	30.6	352	24,038	68.3	188	27,900	148.4
2008	752	23,525	31.3	320	22,297	69.7	200	29,533	147.7
2010	585	18,000	30.8	254	17,203	67.7	130	18,924	145.6
2012	609	18,943	31.1	248	16,983	68.5	143	21,428	149.8

	Size Code 7 250–499			Size Code 8 500–999			Size Code 9 1000 or more		
Year	# of Estab.	Employees	Emp/Est	# of Estab.	Employees	Emp/Est	# of Estab.	Employees	Emp/Est
1990	78	26,915	345.1	25	15,927	637.1	10	16,317	1,631.7
1995	82	28,347	345.7	30	20,168	672.3	6	17,625	2,937.6
2000	88	29,188	331.7	30	19,266	642.2	11	24,525	2,229.6
2005	61	20,235	331.7	13	8,299	638.4	6	12,777	2,129.6
2008	50	16,558	331.2	15	10,304	686.9	6	12,209	2,034.9
2010	30	9,876	329.2	9	5,587	620.8	5*	NA	NA
2012	35	11,436	326.7	7	4,533	647.6	5	9,082	1,816.5

Source: Bureau of Labor Statistics (2013).

1. * Establishment totals for time period had a quarterly status code of "N." Therefore, the Bureau of Labor Statistics includes establishment numbers in the report totals but employment figures are not and subsequently marked as "0" in the employment totals. Information regarding status/disclosure codes is located at the BLS website ftp://ftp.bls.gov/pub/special.requests/cew/DOCUMENT/layout.txt.

2. An "NA" appearing in the number of establishments column for a specific size code denotes that the size code did not appear on the BLS report.

Table 5.11b *Office furniture and fixtures industry change in employment 1990–2012*

Size Code	Change in Employment Due to Change in Establishments			Size Code	Change in Employment Due to Change in Emp/Est			Size Code	Change in Employment Due to Combined Effect			Size Code	Change in Total Employment
	(Δ Est)	(EPE.)	=		(Δ EPE)	(Est.)	=		(Δ Est)	(Δ EPE)	=		
1	234	-2.1	493	1	-0.4	1,206	-477	1	234	-0.4	-93	1	-76
2	-148	6.7	-997	2	-0.1	756	-44	2	-148	-0.1	9	2	-1,032
3	-243	13.8	-3,341	3	0.2	834	139	3	-243	0.2	-41	3	-3,243
4	-225	31.3	-7,044	4	-0.2	834	-167	4	-225	-0.2	45	4	-7,166
5	-126	69.3	-8,735	5	-0.8	374	-314	5	-126	-0.8	106	5	-8,943
6	-142	152.1	-21,599	6	-2.3	285	-644	6	-142	-2.3	321	6	-21,922
7	-43	345.1	-14,838	7	-18.3	78	-1,429	7	-43	-18.3	788	7	-15,479
8	-18	637.1	-11,468	8	10.5	25	262	8	-18	10.5	-189	8	-11,394
9	-5	1,631.7	-8,159	9	184.8	10	1,848	9	-5	184.8	-924	9	-7,235

TOTAL LOSS OF EMPLOYMENT   -76,489

100 to 249 employees, the number of establishments dropped from 514 to 280. As with household furniture, this industry has been hard hit at every size code. The largest loss of jobs was in size code 6, with 100 to 249 employees, as may be seen in Table 5.11b.

For durable goods manufacturing industries, it has been shown that for most industries, the greatest loss of jobs occurred with the closing of larger plants. This seems understandable because it makes more financial sense to off-shore a 1,000-employee plant than a 100-person plant due to economies of scale.

Establishment Closings in the Manufacture of Nondurable Goods

As seen in earlier chapters, several subsectors of the nondurable goods industry sector have experienced enormous losses of both jobs and establishments. It has been shown in the previous section that for durable manufacturing, the loss of jobs was primarily associated with the closings of large size establishments. In this section, the loss of establishments by size code is analyzed for industries of the nondurable goods manufacturing subsector.

Table 5.12a presents the distribution of establishments by size code from 1990 to 2012 for the textile mills industry. The data show an industry being hollowed out from the top down. There are massive losses of establishments and employees at the five largest size codes. Between 1990 and 2012, for size code 9, the drop was from 58 to 4 establishments; for size code 8, the drop was from 151 establishments to 13; for size code 7, the drop was from 365 to 66 establishments; for size code 6, the decline was from 596 to 262; and for size code 5, the drop was from 567 to 264. These closings resulted in job losses totaling 367,000. Of these losses, 288,667 occurred at the largest size codes 7, 8, and 9. Nevertheless, the loss of establishments and jobs appears to penetrate much more deeply into the lower size codes for this industry than those in the durable goods sector. As seen in Table 5.12b, the loss of jobs arises much more from the loss of establishments rather than fewer employees per establishment. For most size codes, the average number of employees per establishment was very stable over the 22-year period. Except for the largest size code, most

Table 5.12a Distribution of establishment size for textile mills industry

	Size Code 1 <5			Size Code 2 5–9			Size Code 3 10–19		
Year	# of Estab.	Employees	Emp/Est	# of Estab.	Employees	Emp/Est	# of Estab.	Employees	Emp/Est
1990	1,982	3,701	1.9	795	5,437	6.8	741	10,202	13.8
1995	2,165	3,952	1.8	799	5,338	6.7	766	10,496	13.7
2000	2,313	3,809	1.6	802	5,444	6.8	689	9,408	13.7
2005	1,457	2,577	1.8	645	4,378	6.8	540	7,613	14.1
2008	1370*	NA	NA	548	3,726	6.8	436	6,017	13.8
2010	1304*	NA	NA	462	3,075	6.7	409	5,705	13.9
2012	1,173	2,123	1.8	424	2,814	6.6	410	5,698	13.9

	Size Code 4 20–49			Size Code 5 50–99			Size Code 6 100–249		
Year	# of Estab.	Employees	Emp/Est	# of Estab.	Employees	Emp/Est	# of Estab.	Employees	Emp/Est
1990	751	23,829	31.7	567	40,401	71.3	596	96,980	162.7
1995	849	27,252	32.1	566	40,293	71.2	654	103,427	158.1
2000	819	25,632	31.3	544	38,228	70.3	593	94,578	159.5

Year	# of Estab.	Employees	Emp/Est	# of Estab.	Employees	Emp/Est	# of Estab.	Employees	Emp/Est
2005	641	20,405	31.8	397	28,484	71.7	400	62,752	156.9
2008	558	17,830	32.0	308	21,929	71.2	331	50,378	152.2
2010	476	15,072	31.7	277	19,599	70.8	260	38,926	149.7
2012	435	13,840	31.8	264	18,567	70.3	262	40,270	153.7

Year	Size Code 7 250–499			Size Code 8 500–999			Size Code 9 1000 or more		
	# of Estab.	Employees	Emp/Est	# of Estab.	Employees	Emp/Est	# of Estab.	Employees	Emp/Est
1990	365	126,615	346.9	151	100,243	663.9	58	96,797	1,668.9
1995	364	125,681	345.3	137	90,793	662.7	39	62,251	1,596.2
2000	294	101,786	346.2	99	67,146	678.2	23	36,540	1,588.7
2005	166	56,703	341.6	46	28,342	616.1	9	12,610	1,401.1
2008	103	35,818	347.7	25	16,029	641.2	4*	NA	NA
2010	65	21,756	334.7	15	9,799	653.2	2*	NA	NA
2012	66	22,231	336.8	13	8,223	632.5	4	4,534	1,133.4

Source: Bureau of Labor Statistics (2013).

1. * Establishment totals for time period had a quarterly status code of "N." Therefore, the Bureau of Labor Statistics includes establishment numbers in the report totals but employment figures are not and subsequently marked as "0" in the employment totals. Information regarding status/disclosure codes is located at the BLS website ftp://ftp.bls.gov/pub/special.requests/cew/DOCUMENT/layout.txt.

2. An "NA" appearing in the number of establishments column for a specific size code denotes that the size code did not appear on the BLS report.

Table 5.12b Textile mills industry change in employment 1990–2012

Size Code	Change in Employment Due to Change in Establishments			Size Code	Change in Employment Due to Change in Emp/Est			Size Code	Change in Employment Due to Combined Effect			Size Code	Change in Total Employment
	(Δ Est)	(EPE.)	=		(Δ EPE)	(Est.)	=		(Δ Est)	(Δ EPE)	=		
1	-809	1.9	-1,511	1	-0.1	1,982	-114	1	-809	-0.1	47	1	-1,578
2	-371	6.8	-2,537	2	-0.2	795	-160	2	-371	-0.2	75	2	-2,623
3	-331	13.8	-4,557	3	0.1	741	97	3	-331	0.1	-43	3	-4,503
4	-316	31.7	-10,027	4	0.1	751	65	4	-316	0.1	-27	4	-9,989
5	-303	71.3	-21,590	5	-0.9	567	-525	5	-303	-0.9	280	5	-21,834
6	-334	162.7	-54,348	6	-9.0	596	-5,374	6	-334	-9.0	3,012	6	-56,710
7	-299	346.9	-103,720	7	-10.1	365	-3,671	7	-299	-10.1	3,007	7	-104,384
8	-138	663.9	-91,613	8	-31.3	151	-4,729	8	-138	-31.3	4,322	8	-92,020
9	-54	1,668.9	-90,121	9	-535.5	58	-31,059	9	-54	-535.5	28,917	9	-92,263

TOTAL LOSS OF EMPLOYMENT   -385,904

of the losses occurred after the year 2000. In summary, this industry lost 49 percent of its establishments, 2,995 businesses closed, and 386,000 jobs were lost. Table 5.12b shows that size codes 6, 7, 8, and 9 had significantly higher job losses than the smaller size codes.

Table 5.13a presents the loss of establishments for the textile product mill industry. This industry lost 17 percent of its plants, 1,448 establishments. Once again the dominant losses were at the four largest size codes. Between 1990 and 2012, for size code 9, the drop was from 12 to 3 establishments; for size code 8, the drop was from 51 establishments to 13; for size code 7, the drop was from 112 to 42 establishments; for size code 6, the decline was from 306 to 170; and for size code 5, the drop was from 406 to 232. As in the case for textile mills, the loss of establishments and jobs appears to penetrate much more deeply into the lower size codes for this industry. And again, as seen in Table 5.13b, the loss of jobs arises much more from the loss of establishments rather than fewer employees per establishment. For most codes, the average number of employees per establishment was very stable over the 22-year period. Except for the largest size code, most of the losses occurred after the year 2000.

Table 5.14a presents the loss of establishments for the apparel industry. This industry has been decimated at just about every size code, losing 7,272 establishments and over 750,000 employees. This industry has been concentrated in the small to medium size establishments with relatively few plants with over a thousand workers, that is, size code 9. Between 1990 and 2012, for size code 8, the drop was from 168 to 18 establishments; for size code 7, the drop was from 568 establishments to 60; for size code 6, the drop was from 1,655 to just 222 establishments; for size code 5, the decline was from 2,142 to 322, and for size code 4, the drop was from 3,962 to 818. Even size code 3 with 10 to 19 employees had the number of establishments drop from 3,009 to 1,085. As opposed to many of the other industries, the losses in this industry hit every size code. Again, as shown in Table 5.14b, the average number of employees per establishment remained relatively stable—the job losses were dominantly associated with the closing of apparel factories.

In nearly all industries, larger facilities are typically more efficient and lower cost because they can allocate fixed costs over a greater number of output units. This certainly has been the case in agriculture where the

Table 5.13a Distribution of establishment size for textile product mills industry

Year	Size Code 1 < 5			Size Code 2 5–9			Size Code 3 10–19		
	# of Estab.	Employees	Emp/Est	# of Estab.	Employees	Emp/Est	# of Estab.	Employees	Emp/Est
1990	4,064	7,678	1.9	1,609	10,455	6.5	1,152	15,601	13.5
1995	4,261	7,756	1.8	1,619	10,556	6.5	1,132	15,194	13.4
2000	4,216	7,418	1.8	1,654	10,888	6.6	1,119	14,973	13.4
2005	3758*	NA	NA	1,455	9,488	6.5	1,010	13,559	13.4
2008	4,246	7,513	1.8	1,516	9,877	6.5	1,022	13,839	13.5
2010	4,353	7,268	1.7	1,351	8,861	6.6	815	10,856	13.3
2012	3997*	NA	NA	1,285	8,305	6.5	852	11,399	13.4

Year	Size Code 4 20–49			Size Code 5 50–99			Size Code 6 100–249		
	# of Estab.	Employees	Emp/Est	# of Estab.	Employees	Emp/Est	# of Estab.	Employees	Emp/Est
1990	919	27,934	30.4	406	28,337	69.8	306	47,123	154.0
1995	917	27,397	29.9	416	29,517	71.0	313	48,254	154.2
2000	928	28,102	30.3	388	27,025	69.7	300	45,180	150.6

	774	23,287	30.1	305	20,682	67.8	220	33,918	154.2
2005	774	23,287	30.1	305	20,682	67.8	220	33,918	154.2
2008	715	22,267	31.1	294	20,952	71.3	206	32,138	156.0
2010	600	18,302	30.5	226	16,227	71.8	172	25,612	148.9
2012	589	18,083	30.7	232	16,367	70.5	170	26,564	156.3

	Size Code 7 250–499			Size Code 8 500–999			Size Code 9 1000 or more		
Year	# of Estab.	Employees	Emp/Est	# of Estab.	Employees	Emp/Est	# of Estab.	Employees	Emp/Est
1990	112	37,863	338.1	51	33,654	659.9	12	17,993	1,499.4
1995	112	39,342	351.3	49	31,754	648.0	12	16,413	1,367.8
2000	117	40,891	349.5	43	29,517	686.4	9	13,217	1,468.6
2005	96	33,354	347.4	33	21,942	664.9	4*	NA	NA
2008	73	24,668	337.9	20*	NA	NA	4*	NA	NA
2010	50	16,644	332.9	14	9,016	644.0	3	4,905	1,635.0
2012	42	14,747	351.1	13	8,412	647.1	3*	NA	NA

Source: Bureau of Labor Statistics (2013).

1. * Establishment totals for time period had a quarterly status code of "N." Therefore, the Bureau of Labor Statistics includes establishment numbers in the report totals but employment figures are not and subsequently marked as "0" in the employment totals. Information regarding status/disclosure codes is located at the BLS website ftp://ftp.bls.gov/pub/special.requests/cew/DOCUMENT/layout.txt.

2. An "NA" appearing in the number of establishments column for a specific size code denotes that the size code did not appear on the BLS report.

Table 5.13b Textile product mills industry change in employment 1990–2012

Size Code	Change in Employment Due to Change in Establishments			Size Code	Change in Employment Due to Change in Emp/Est			Size Code	Change in Employment Due to Combined Effect			Size Code	Change in Total Employment
	(Δ Est)	(EPE.)	=		(Δ EPE)	(Est.)	=		(Δ Est)	(Δ EPE)	=		
1	NA	NA	NA	1	NA	NA	NA	1	NA	NA	NA	1	NA
2	-324	6.5	-2,105	2	0.0	1,609	-56	2	-324	0.0	11	2	-2,150
3	-300	13.5	-4,063	3	-0.2	1,152	-188	3	-300	-0.2	49	3	-4,202
4	-330	30.4	-10,031	4	0.3	919	280	4	-330	0.3	-100	4	-9,852
5	-174	69.8	-12,145	5	0.7	406	304	5	-174	0.7	-130	5	-11,971
6	-136	154.0	-20,944	6	2.3	306	692	6	-136	2.3	-307	6	-20,559
7	-70	338.1	-23,665	7	13.0	112	1,461	7	-70	13.0	-913	7	-23,117
8	-38	659.9	-25,075	8	-12.8	51	-651	8	-38	-12.8	485	8	-25,241
9	NA	NA	NA	9	NA	NA	NA	9	NA	NA	NA	9	NA

TOTAL LOSS OF EMPLOYMENT   -97,092

Table 5.14a Distribution of establishment size for apparel industry

Year	Size Code 1 < 5 # of Estab.	Employees	Emp/Est	Size Code 2 5–9 # of Estab.	Employees	Emp/Est	Size Code 3 10–19 # of Estab.	Employees	Emp/Est
1990	NA	NA	NA	2,972	19,746	6.6	3,009	40,930	13.6
1995	NA	NA	NA	3,078	20,144	6.5	3,228	43,194	13.4
2000	NA	NA	NA	2,711	17,983	6.6	2,861	38,373	13.4
2005	4,863	9,032	1.9	1,959	13,102	6.7	1,840	24,868	13.5
2008	4,007	7,362	1.8	1,661	10,783	6.5	1,397	18,707	13.4
2010	3,720	6,322	1.7	1,412	9,044	6.4	1,134	14,987	13.2
2012	3,377	5,808	1.7	1,299	8,338	6.4	1,085	14,447	13.3

Year	Size Code 4 20–49 # of Estab.	Employees	Emp/Est	Size Code 5 50–99 # of Estab.	Employees	Emp/Est	Size Code 6 100–249 # of Estab.	Employees	Emp/Est
1990	3,962	121,520	30.7	2,142	149,946	70.0	1,655	252,970	152.9
1995	3,559	109,313	30.7	1,870	128,298	68.6	1,370	210,913	154.0
2000	2,892	87,097	30.1	1,234	83,858	68.0	780	119,381	153.1
2005	1,501	45,164	30.1	618	42,385	68.6	388	57,632	148.5
2008	1,173	35,043	29.9	461	32,005	69.4	301	45,675	151.7

(Continued)

Table 5.14a (Continued)

	Size Code 4 20–49			Size Code 5 50–99			Size Code 6 100–249		
Year	# of Estab.	Employees	Emp/Est	# of Estab.	Employees	Emp/Est	# of Estab.	Employees	Emp/Est
2010	867	26,099	30.1	359	24,833	69.2	232	35,581	153.4
2012	818	24,255	29.7	322	21,953	68.2	222	33,035	148.8

	Size Code 7 250–499			Size Code 8 500–999			Size Code 9 1000 or more		
Year	# of Estab.	Employees	Emp/Est	# of Estab.	Employees	Emp/Est	# of Estab.	Employees	Emp/Est
1990	568	195,580	344.3	168	111,006	660.8	NA	NA	NA
1995	447	154,078	344.7	180	118,337	657.4	NA	NA	NA
2000	223	76,096	341.2	93	62,187	668.7	NA	NA	NA
2005	102	35,861	351.6	35	23,186	662.5	6	10,293	1,715.6
2008	74	24,662	333.3	27	17,664	654.2	5	9,883	1,976.7
2010	56	19,080	340.7	17	10,707	629.8	4	9,826	2,456.6
2012	60	19,956	332.6	18	11,924	662.5	3	8,037	2,678.9

Source: Bureau of Labor Statistics (2013).

1. * Establishment totals for time period had a quarterly status code of "N." Therefore, the Bureau of Labor Statistics includes establishment numbers in the report totals but employment figures are not and subsequently marked as "0" in the employment totals. Information regarding status/disclosure codes is located at the BLS website ftp://ftp.bls.gov/pub/special.requests/cew/DOCUMENT/layout.txt.

2. An "NA" appearing in the number of establishments column for a specific size code denotes that the size code did not appear on the BLS report.

Table 5.14b *Apparel industry change in employment 1990–2012*

Size Code	Change in Employment Due to Change in Establishments			Size Code	Change in Employment Due to Change in Emp/Est			Size Code	Change in Employment Due to Combined Effect			Size Code	Change in Total Employment
	(ΔEst)	(EPE.)	=		(ΔEPE)	(Est.)	=		(ΔEst)	(ΔEPE)	=		
1	NA	NA	NA	1	NA	NA	NA	1	NA	NA	NA	1	NA
2	-1,673	6.6	-11,116	2	-0.2	2,972	-669	2	-1,673	-0.2	377	2	-11,408
3	-1,924	13.6	-26,171	3	-0.3	3,009	-865	3	-1,924	-0.3	553	3	-26,483
4	-3,144	30.7	-96,431	4	-1.0	3,962	-4,042	4	-3,144	-1.0	3,207	4	-97,265
5	-1,820	70.0	-127,405	5	-1.8	2,142	-3,908	5	-1,820	-1.8	3,321	5	-127,992
6	-1,433	152.9	-219,037	6	-4.0	1,655	-6,695	6	-1,433	-4.0	5,797	6	-219,935
7	-508	344.3	-174,920	7	-11.7	568	-6,666	7	-508	-11.7	5,962	7	-175,624
8	-150	660.8	-99,113	8	1.7	168	287	8	-150	1.7	-257	8	-99,082
9	NA	NA	NA	9	NA	NA	NA	9	NA	NA	NA	9	NA

TOTAL LOSS OF EMPLOYMENT   -757,790

number of farms has fallen and the average size has increased. This trend achieves greater efficiency and productivity. For the textile and apparel industries, however, the opposite is seen to have occurred—larger firms have tended to disappear. Those plants, of course, would be the easiest to off-shore and, because of the volume, make the most financial sense to send off-shore.

Table 5.15a presents the loss of establishments for the paper and paper product industry. Once again, the impacts in this industry are concentrated primarily in the larger size codes. Between 1990 and 2012, for size code 9, the drop was from 86 to 21 establishments with a loss of 109,000 jobs; for size code 8, the drop was from 151 establishments to 73 with a loss of 51,000 jobs; for size code 7, the drop was from 306 to 191 establishments with a loss of 36,000 jobs; for size code 6, the decline was from 1,156 to 826 with 51,000 jobs; and for size code 5, the drop was from 1,035 to 850 with the loss of 12,000 jobs. For the paper industry, the loss of establishments and jobs is much more concentrated in the larger size codes. And again, as seen in Table 5.15b, the loss of jobs arises much more from the loss of establishments rather than fewer employees per establishment. For most codes, the average number of employees per establishment was very stable over the 22-year period, it is the drop in the number of establishments that drives the job losses, a total of 271,000.

The printing industry has been hard hit with job and establishment losses. Table 5.16a presents the loss of establishments for this industry. Significant losses have occurred for every establishment size code. Between 1990 and 2012, for size code 9, the drop was from 24 to 5 establishments; for size code 8, the drop was from 75 establishments to 45; for size code 7, the drop was from 290 to 146 establishments; for size code 6, the decline was from 1,061 to 663; and for size code 5, the drop was from 1,969 to 1,068. Even the smaller establishment size codes were hard hit in the printing industry. For size code 4 (20 to 49 employees), the drop was from 4,893 to 2,823; for size code 3 (10 to 19 employees), the drop was from 7,100 to 3,671; for size code 2 (5 to 9 employees), the drop was from 10,705 to 5,706; and for size code 1 (less than 5 employees), the drop was from 22,219 to 16,380 establishments. Significant job losses occurred at every size code as may be seen in Table 5.16b. Total jobs lost between 1990 and 2012 were 373,693. As will be seen in a later

Table 5.15a Distribution of establishment size for paper and paper products industry

	Size Code 1 < 5			Size Code 2 5–9			Size Code 3 10–19		
Year	# of Estab.	Employees	Emp/Est	# of Estab.	Employees	Emp/Est	# of Estab.	Employees	Emp/Est
1990	900	1,781	2.0	614	4,226	6.9	912	12,981	14.2
1995	1,004	1,881	1.9	623	4,274	6.9	867	12,337	14.2
2000	1,244	2,225	1.8	634	4,334	6.8	868	12,376	14.3
2005	1,329	2,712	2.0	618	4,204	6.8	785	11,029	14.0
2008	1,401	4,218	3.0	623	4,224	6.8	763	10,799	14.2
2010	1,451	2,621	1.8	613	4,410	7.2	726	10,275	14.2
2012	1,348	2,591	1.9	579	3,999	6.9	677	9,565	14.1

	Size Code 4 20–49			Size Code 5 50–99			Size Code 6 100–249		
Year	# of Estab.	Employees	Emp/Est	# of Estab.	Employees	Emp/Est	# of Estab.	Employees	Emp/Est
1990	1,385	45,371	32.8	1,035	73,935	71.4	1,156	173,859	150.4
1995	1,397	46,165	33.0	1,118	79,974	71.5	1,255	192,124	153.1
2000	1,432	46,979	32.8	1,140	81,651	71.6	1,266	192,230	151.8

(Continued)

Table 5.15a (Continued)

Year	Size Code 4 20–49			Size Code 5 50–99			Size Code 6 100–249		
	# of Estab.	Employees	Emp/Est	# of Estab.	Employees	Emp/Est	# of Estab.	Employees	Emp/Est
2005	1,299	42,664	32.8	1,026	73,320	71.5	1,075	163,462	152.1
2008	1,210	39,815	32.9	975	69,944	71.7	978	146,411	149.7
2010	1,150	37,433	32.6	887	63,569	71.7	840	123,425	146.9
2012	1,132	36,918	32.6	850	61,283	72.1	826	122,600	148.4

Year	Size Code 7 250–499			Size Code 8 500–999			Size Code 9 1000 or more		
	# of Estab.	Employees	Emp/Est	# of Estab.	Employees	Emp/Est	# of Estab.	Employees	Emp/Est
1990	306	102,461	334.8	151	100,602	666.2	86	135,921	1,580.5
1995	286	97,775	341.9	143	97,168	679.5	74	114,145	1,542.5
2000	278	94,227	338.9	132	90,478	685.4	55	84,768	1,541.2
2005	211	72,589	344.0	103	69,509	674.8	36	48,024	1,334.0
2008	209	71,155	340.5	96	63,824	664.8	28	37,172	1,327.6
2010	205	71,446	348.5	77	54,875	712.7	19	24,897	1,310.4
2012	191	66,500	348.2	73	49,829	682.6	21	26,855	1,278.8

Source: Bureau of Labor Statistics (2013).

Table 5.15b Paper and paper products industry change in employment 1990–2012

Size Code	Change in Employment Due to Change in Establishments			Size Code	Change in Employment Due to Change in Emp/Est			Size Code	Change in Employment Due to Combined Effect			Size Code	Change in Total Employment
	(Δ Est)	(EPE.)	=		(Δ EPE)	(Est.)	=		(Δ Est)	(Δ EPE)	=		
1	448	2.0	887	1	-0.1	900	-51	1	448	-0.1	-26	1	810
2	-35	6.9	-241	2	0.0	614	14	2	-35	0.0	-1	2	-227
3	-235	14.2	-3,345	3	-0.1	912	-96	3	-235	-0.1	25	3	-3,416
4	-253	32.8	-8,288	4	-0.1	1,385	-202	4	-253	-0.1	37	4	-8,453
5	-185	71.4	-13,215	5	0.7	1,035	686	5	-185	0.7	-123	5	-12,652
6	-330	150.4	-49,631	6	-2.0	1,156	-2,278	6	-330	-2.0	650	6	-51,258
7	-115	334.8	-38,507	7	13.3	306	4,077	7	-115	13.3	-1,532	7	-35,962
8	-78	666.2	-51,967	8	16.3	151	2,469	8	-78	16.3	-1,275	8	-50,773
9	-65	1,580.5	-102,731	9	-301.7	86	-25,944	9	-65	-301.7	19,609	9	-109,066

TOTAL LOSS OF EMPLOYMENT   -270,999

Table 5.16a Distribution of establishment size for printing and related support activities industry

Year	Size Code 1 < 5			Size Code 2 5–9			Size Code 3 10–19		
	# of Estab.	Employees	Emp/Est	# of Estab.	Employees	Emp/Est	# of Estab.	Employees	Emp/Est
1990	22,219	46,481	2.1	10,705	69,520	6.5	7,100	94,478	13.3
1995	22,568	45,330	2.0	10,398	67,950	6.5	6,726	89,773	13.3
2000	21,043	41,192	2.0	9,327	60,921	6.5	6,235	84,045	13.5
2005	18,545	36,524	2.0	7,676	50,551	6.6	5,112	68,642	13.4
2008	17461*	NA	NA	7,111	46,709	6.6	4,620	62,089	13.4
2010	17,392	32,067	1.8	6,141	40,199	6.5	3,826	51,769	13.5
2012	16,380	30,424	1.9	5,706	37,101	6.5	3,671	49,094	13.4

Year	Size Code 4 20–49			Size Code 5 50–99			Size Code 6 100–249		
	# of Estab.	Employees	Emp/Est	# of Estab.	Employees	Emp/Est	# of Estab.	Employees	Emp/Est
1990	4,893	148,182	30.3	1,969	135,095	68.6	1,061	155,879	146.9
1995	4,800	146,362	30.5	1,872	129,517	69.2	1,073	159,843	149.0
2000	4,563	138,664	30.4	1,836	127,382	69.4	1,110	165,421	149.0

Year	# of Estab.	Employees	Emp/Est	# of Estab.	Employees	Emp/Est	# of Estab.	Employees	Emp/Est
2005	3,745	114,133	30.5	1,473	100,677	68.3	886	132,370	149.4
2008	3,561	109,618	30.8	1,442	99,269	68.8	862	129,109	149.8
2010	2,994	92,299	30.8	1,157	80,428	69.5	682	102,030	149.6
2012	2,823	86,787	30.7	1,068	73,722	69.0	663	97,751	147.4

	Size Code 7 250–499			Size Code 8 500–999			Size Code 9 1000 or more		
Year	# of Estab.	Employees	Emp/Est	# of Estab.	Employees	Emp/Est	# of Estab.	Employees	Emp/Est
1990	290	97,240	335.3	75	50,542	673.9	24	36,685	1,528.5
1995	279	93,017	333.4	86	57,577	669.5	21	32,993	1,571.1
2000	278	94,895	341.3	102	68,946	675.9	20	28,844	1,442.2
2005	225	75,832	337.0	69	47,455	687.8	15	18,811	1,254.1
2008	198	66,776	337.3	64	42,993	671.8	12*	NA	NA
2010	143	48,737	340.8	49	32,631	665.9	6	7,070	1,178.3
2012	146	48,072	329.3	45	30,502	677.8	5	6,957	1,391.3

Source: Bureau of Labor Statistics (2013).

1. * Establishment totals for time period had a quarterly status code of "N." Therefore, the Bureau of Labor Statistics includes establishment numbers in the report totals but employment figures are not and subsequently marked as "0" in the employment totals. Information regarding status/disclosure codes is located at the BLS website ftp://ftp.bls.gov/pub/special.requests/cew/DOCUMENT/layout.txt.

2. An "NA" appearing in the number of establishments column for a specific size code denotes that the size code did not appear on the BLS report.

Table 5.16b Printing and related support activities industry change in employment 1990–2012

Size Code	Change in Employment Due to Change in Establishments			Size Code	Change in Employment Due to Change in Emp/Est			Size Code	Change in Employment Due to Combined Effect			Size Code	Change in Total Employment
	(Δ Est)	(EPE.)	=		(Δ EPE)	(Est.)	=		(Δ Est)	(Δ EPE)	=		
1	-5,839	2.1	-12,215	1	-0.2	22,219	-5,213	1	-5,839	-0.2	1,370	1	-16,058
2	-4,999	6.5	-32,464	2	0.0	10,705	86	2	-4,999	0.0	-40	2	-32,418
3	-3,429	13.3	-45,629	3	0.1	7,100	473	3	-3,429	0.1	-229	3	-45,384
4	-2,070	30.3	-62,689	4	0.5	4,893	2,243	4	-2,070	0.5	-949	4	-61,395
5	-901	68.6	-61,818	5	0.4	1,969	822	5	-901	0.4	-376	5	-61,373
6	-398	146.9	-58,473	6	0.5	1,061	552	6	-398	0.5	-207	6	-58,128
7	-144	335.3	-48,285	7	-6.1	290	-1,755	7	-144	-6.1	871	7	-49,168
8	-30	673.9	-20,217	8	3.9	75	294	8	-30	3.9	-118	8	-20,040
9	-19	1,528.5	-29,042	9	-137.2	24	-3,293	9	-19	-137.2	2,607	9	-29,728

TOTAL LOSS OF EMPLOYMENT   -373,693

chapter, this arises from the use of high quality printers in the home and office and from the decline in paper publications such as newspapers and magazines. Total job losses in the printing industry from 1990 to 2012 were 374,000.

Table 5.17 presents an overview of the number of establishments for the previously discussed nondurable industries. As may be seen, textile mills and apparel both lost roughly 50 percent of establishments with over 10,000 plants closing. Printing lost nearly 18,000 establishments or 37 percent of the 1990 number. Textile mills and paper industries lost 17 percent and 13 percent of establishments, respectively. For these industries, the following observations may be made:

1. Between 1990 and 2012, the number of establishments declined for *every* size code;
2. Over that period, the total number of employees at every size code went down;
3. The average number of employees per establishment in each size code remained very stable; and
4. In each industry, the percentage loss of large establishments was much greater than the loss of medium and small establishments.

Table 5.17 Total number of establishments by year

Year	Textile Mills	Textile Product Mills	Apparel	Paper & Paper Products	Printing & Related Support Activities
1990	6,006	8,631	14,476	6,545	48,336
1995	6,339	8,831	13,732	6,767	47,823
2000	6,176	8,774	10,794	7,049	44,514
2005	4,301	7,655	11,312	6,482	37,746
2008	3,683	8,096	9,106	6,283	35,331
2010	3,270	7,584	7,801	5,968	32,390
2012	3,051	7,183	7,204	5,697	30,507
Number Closed	2,955	1,448	7,272	848	17,829
% Closed	−49.2%	−16.8%	−50.2%	−13.0%	−36.9%

Source: Bureau of Labor Statistics (2013).

These are not trends one would expect to see if the losses were arising dominantly from growth in productivity. One would expect to see the emergence of more large firms rather than smaller establishments.

Summary

The movement toward larger firms is, in fact, exactly what happened in agriculture. Figure 5.2 presents long-term trends for the number of farms, the average size of farms, and the volume of land in farming. As may be seen, the number of farms has declined sharply and the average size has more than doubled. The land in farming has remained fairly stable, with just a slight decline. Since the loss of jobs in manufacturing has been compared to the evolution of the agricultural sector, farm output and productivity are discussed in detail in Chapter 6.

Table 5.18 presents the loss of establishments by size code for all of manufacturing. The figures in this Figure are rather astounding. Between 1990 and 2000, for all of U.S. manufacturing, the number of establishments was stable or slightly growing at every size code but one. Over the same time period, total employment was stable or slightly growing at every size code but one. The single exception was in size code 9 (over 1,000 employees) where between 1990 and 2000, the number of establishments went from 1,705 to 1,514 and total employment at that size code dropped from 4.1 million to 3.2 million. Most of that drop occurred in just two industries: The aerospace products and parts industry and the

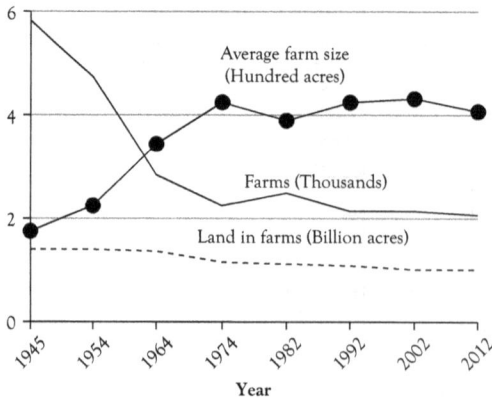

Figure 5.2 Farms, average farm size, and land in farms
Source: USDA (2012).

Table 5.18 Distribution of establishment size for U.S. manufacturing

Year	Size Code 1 <5			Size Code 2 5–9			Size Code 3 10–19		
	# of Estab.	Employees	Emp/Est	# of Estab.	Employees	Emp/Est	# of Estab.	Employees	Emp/Est
1990	134,731	266,629	2.0	68,480	487,633	7.1	61,333	834,754	13.6
1995	144,084	274,834	1.9	69,629	462,075	6.6	62,282	845,652	13.6
2000	150,808	273,116	1.8	68,401	455,116	6.7	61,413	838,228	13.6
2005	139,265	261,576	1.9	62,539	416,779	6.7	55,531	756,825	13.6
2008	138,761	265,611	1.9	61,564	412,030	6.7	53,932	739,489	13.7
2010	144,457	255,053	1.8	57,867	384,878	6.7	49,622	678,242	13.7
2012	135,937	242,746	1.8	56,151	371,841	6.6	49,032	667,918	13.6

Year	Size Code 4 20–49			Size Code 5 50–99			Size Code 6 100–249		
	# of Estab.	Employees	Emp/Est	# of Estab.	Employees	Emp/Est	# of Estab.	Employees	Emp/Est
1990	58,600	1,808,569	30.9	28,325	1,978,742	69.9	21,955	3,370,020	153.5
1995	59,870	1,847,027	30.9	28,826	2,006,759	69.6	22,201	3,408,107	153.5
2000	60,062	1,859,261	31.0	29,394	2,049,280	69.7	22,990	3,529,313	153.5

(Continued)

Table 5.18 (Continued)

Year	Size Code 4 20–49			Size Code 5 50–99			Size Code 6 100–249		
	# of Estab.	Employees	Emp/Est	# of Estab.	Employees	Emp/Est	# of Estab.	Employees	Emp/Est
2005	53,217	1,646,314	30.9	25,598	1,784,166	69.7	19,498	2,986,731	153.2
2008	52,329	1,629,805	31.1	25,129	1,756,202	69.9	18,998	2,910,968	153.2
2010	46,322	1,433,885	31.0	21,540	1,496,739	69.5	15,696	2,375,848	151.4
2012	46,687	1,442,641	30.9	22,305	1,550,577	69.5	16,555	2,513,242	151.8

Year	Size Code 7 250–499			Size Code 8 500–999			Size Code 9 1000 or more		
	# of Estab.	Employees	Emp/Est	# of Estab.	Employees	Emp/Est	# of Estab.	Employees	Emp/Est
1990	7,605	2,615,153	343.9	3,312	2,251,260	679.7	1,705	4,131,419	2,423.1
1995	7,814	2,693,937	344.8	3,292	2,230,074	677.4	1,551	3,331,398	2,147.9
2000	8,046	2,764,869	343.6	3,298	2,238,717	678.8	1,514	3,255,988	2,150.6
2005	6,468	2,227,274	344.4	2,432	1,643,268	675.7	1,155	2,400,451	2,078.3
2008	6,052	2,074,024	342.7	2,298	1,559,671	678.7	1,065	2,216,622	2,081.3
2010	4,830	1,655,341	342.7	1,800	1,210,978	672.8	881	1,838,638	2,087.0
2012	5,132	1,760,897	343.1	1,913	1,279,474	668.8	913	1,921,174	2,104.2

Source: Bureau of Labor Statistics (2013).

search, detection, navigation, guidance, aeronautical, and nautical system and instrument manufacturing industry. Both of these industries were hard hit by the substantial reductions in defense spending following the end of the Cold War as shown in Figure 5.2. Both of these industries went through significant consolidation and shrinkage between 1990 and 2000. Not only did employment and establishments decline in this period, but as seen in Figure 5.3, output, as measured by the Federal Reserve's production index, for the defense and space market group fell dramatically. Again, this is a very different phenomenon from that seen in agriculture where falling of employment and farms was accompanied by higher output. Here, in the aerospace and defense industries, falling employment and establishments followed declining output and sales between 1990 and 2000.

After the year 2000, however, the declines in employment and establishments hit many industries and the bottom dropped out of many manufacturing industries as the number of establishments and the employment plummeted.

Table 5.19 presents an overview summary of the loss of total employment by size code for a number of durable goods manufacturing industries. The losses of employment over 40 percent are highlighted in bold and underlined. As may be readily seen, the greatest loss of employment occurred within the largest size codes with losses ranging from roughly 50–80 percent. In a number of cases, the sizeable losses of employment extend down to size codes as small as 50 to 250 employees. As a result,

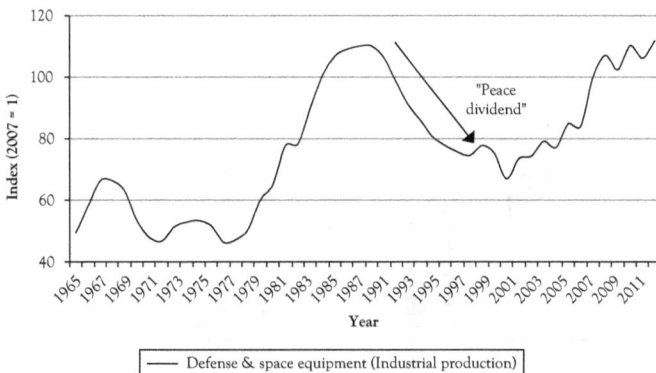

Figure 5.3 Industrial production index for defense and space equipment

Source: The Federal Reserve (2013).

Table 5.19 Percentage change in employment for important durable goods industries, 1990–2012, for different size codes (losses over 40 percent are highlighted in bold and underlined)

	Size Code 1 <5	Size Code 2 5–9	Size Code 3 10–19	Size Code 4 20–49	Size Code 5 50–99	Size Code 6 100–249	Size Code 7 250–499	Size Code 8 500–999	Size Code 9 1000 or more
Motor Vehicle Parts	-0.7%	-17.6%	-29.0%	-24.8%	-17.4%	-10.2%	-1.5%	-8.7%	**-72.2%**
Foundries	-28.5%	-25.5%	-32.4%	**-40.7%**	-32.2%	**-43.1%**	-18.0%	-36.4%	**-77.4%**
Forging & Stamping	0.0%	-23.5%	-18.0%	-13.1%	-9.8%	-9.0%	-6.5%	-31.5%	0.0%
Metalworking Machinery	-25.3%	-36.3%	-38.7%	-27.1%	-23.7%	-20.6%	-37.0%	**-71.1%**	0.0%
Household Appliances	40.1%	58.0%	-3.1%	-7.0%	42.9%	**-40.6%**	**-43.7%**	**-68.8%**	**-49.2%**
Computer & Peripheral Equipment	-6.9%	-28.0%	-27.1%	-31.7%	**-42.7%**	**-41.5%**	**-26.4%**	**-67.7%**	**-54.4%**
Communications Equipment	47.3%	-9.7%	1.4%	-26.0%	-25.7%	-30.5%	-39.9%	**-54.4%**	**-80.0%**
Semiconductors & Electronic Components	5.9%	-2.3%	-7.3%	-10.7%	-22.4%	-22.1%	-33.9%	**-47.3%**	**-38.5%**
Aerospace Products & Parts Industry	55.5%	17.9%	-7.7%	2.5%	0.4%	24.9%	22.4%	27.4%	**-55.0%**
Household & Institutional Furniture	-34.6%	**-41.1%**	-38.1%	**-41.1%**	**-44.1%**	**-47.0%**	**-64.2%**	**-47.9%**	**-69.4%**
Office Furniture & Fixtures	-3.0%	-20.3%	-28.3%	-27.4%	-34.5%	**-50.6%**	**-57.5%**	**-71.5%**	**-44.3%**
Total U.S. Manufacturing	-9.0%	-23.7%	-20.0%	-20.2%	-21.6%	-25.4%	-32.7%	**-43.2%**	**-53.5%**

Source: Bureau of Labor Statistics (2013).

Table 5.20 Percentage change in the number of establishments, 1990–2012, for different size codes and industries (losses over 40 percent are highlighted in bold and underlined)

	Size Code 1 <5	Size Code 2 5–9	Size Code 3 10–19	Size Code 4 20–49	Size Code 5 50–99	Size Code 6 100–249	Size Code 7 250–499	Size Code 8 500–999	Size Code 9 1000 or more
Motor Vehicle Parts	13.1%	−16.7%	−30.2%	−26.0%	−16.7%	−9.4%	−2.6%	−5.3%	**−54.0%**
Foundries	−7.1%	−24.0%	−31.8%	**−43.3%**	−32.2%	**−42.9%**	−19.4%	−38.3%	**−76.2%**
Forging & Stamping	0.0%	−23.5%	−18.7%	−15.0%	−6.7%	−7.1%	0.0%	−30.0%	0.0%
Metalworking Machinery	−19.9%	−35.3%	−38.4%	−26.5%	−22.3%	−19.3%	−33.8%	**−68.0%**	0.0%
Household Appliances	50.8%	57.8%	−6.0%	−2.2%	40.9%	−30.2%	**−40.5%**	**−70.0%**	**−46.4%**
Computer & Peripheral Equipment	26.8%	−25.8%	−25.6%	−31.8%	**−44.2%**	**−44.6%**	−23.0%	**−66.2%**	**−57.6%**
Communications Equipment	63.0%	−11.6%	0.7%	−25.4%	−27.1%	−30.9%	**−40.7%**	**−53.1%**	**−71.7%**
Semiconductors & Electronic Components	13.5%	−6.9%	−8.2%	−12.6%	−22.0%	−20.9%	−31.1%	**−44.2%**	−36.1%
Aerospace Products & Parts Industry	83.2%	19.8%	−10.6%	0.8%	−1.1%	29.9%	21.7%	33.3%	−32.4%
Household & Institutional Furniture	−24.6%	**−40.7%**	−38.0%	**−40.1%**	**−44.8%**	**−45.3%**	**−64.5%**	**−49.3%**	**−71.4%**
Office Furniture & Fixtures	19.4%	−19.6%	−29.1%	−27.0%	−33.7%	**−49.8%**	**−55.1%**	**−72.0%**	**−50.0%**
Total U.S. Manufacturing	0.9%	−18.0%	−20.1%	−20.3%	−21.3%	−24.6%	−32.5%	−39.3%	**−46.5%**

Source: Bureau of Labor Statistics (2013).

employment is now concentrated in small to medium size manufacturing plants. However, the loss of employment for a number of industries extended across nearly every size code down to the smallest size facilities.

Table 5.20 presents an overview summary of the loss of establishments by size code for a number of durable goods manufacturing industries. The losses of establishments over 40 percent are highlighted in bold and underlined. As may be readily seen, the greatest loss of establishments occurred within the largest size codes with losses ranging from roughly 40–80 percent. In a number of cases, the sizeable losses of establishments extend down to size codes as small as 50 to 250 employees. As a result, most establishments are now concentrated in small to medium size manufacturing plants.

Key Takeaways

Between 1990 and 2012, many manufacturing industries lost 50–70 percent of their largest establishments, those employing more than 1,000 employees and those employing between 500 and 999 employees. As a result, American manufacturing is now more concentrated in small to medium size facilities. These smaller facilities tend to have less cash flow, lower profit margins, reduced ability to invest in new technologies, and greater susceptibility to downturns in the economy. This enhances the risk of further contraction of U.S. manufacturing. The fundamental data pattern seen in this chapter is shown.

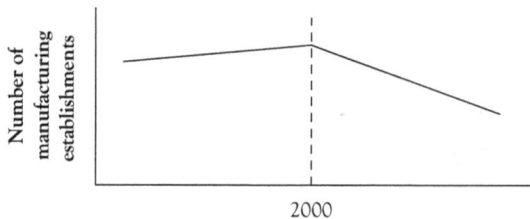

Data pattern 3: Number of large manufacturing establishments

CHAPTER 6

Industrial Production and Industry Output

Is Manufacturing Doing Just Fine? Is it Simply Following the Agricultural Model?

As shown in previous chapters, U.S. manufacturing has lost millions of jobs and tens of thousands of manufacturing establishments. Nevertheless, many claim that all is well with U.S. manufacturing and that output and production are up, comparing the situation with the growth of agricultural output accomplished with many fewer employees. Some examples of this thinking as provided by Atkinson (Atkinson et al. 2012):

> "The long-term trends that we have recently seen in manufacturing mirror what we saw in agriculture a couple of generations ago." N. Gregory Mankiw, Chairman of the Council of Economic Advisers (CEA) under President Bush.

> "Manufacturing employment has fallen (since 2000) because of productivity growth, not a decline in output." The Congressional Research Service.

> "The majority of manufacturing job losses is due to productivity increases." Robert Reich, former secretary of labor, University of California, Berkeley.

> "Manufacturing is doing amazingly well." Mark Perry, American Enterprise Institute.

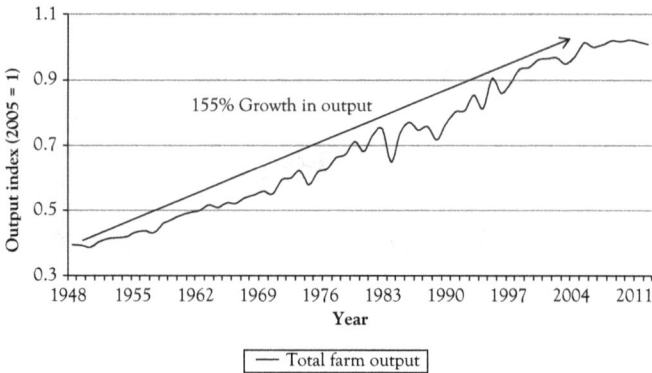

Figure 6.1 Index for total farm output
Source: Fuglie, MacDonald, and Ball (2007).

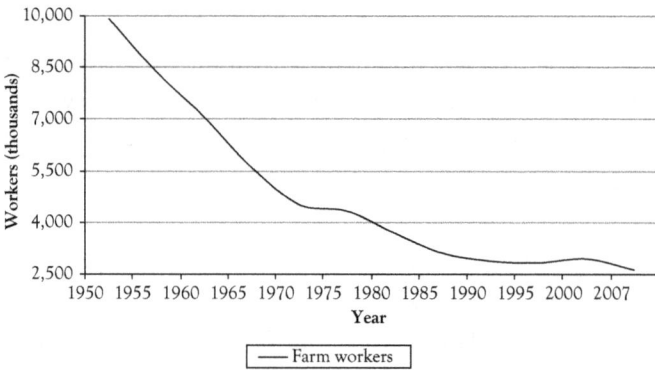

Figure 6.2 Number of U.S. farm workers
Source: U.S. Department of agriculture (n.d.).

Figure 6.1 presents the index of total farm output from the U.S. Department of Agriculture (USDA). Over the period 1948–2011, the index for farm output grew by 156 percent for an annual average cumulative growth rate of 1.50 percent. Over a similar time period from 1948–2007, the number of farm workers declined from 10 million to 2.6 million (Figure 6.2). Thus farms are producing over twice the output with roughly one-fourth the number of workers. This is a remarkable story of productivity growth in the U.S. agricultural sector. Examples of this productivity growth are provided in a report by the Economic Research Service of the USDA:

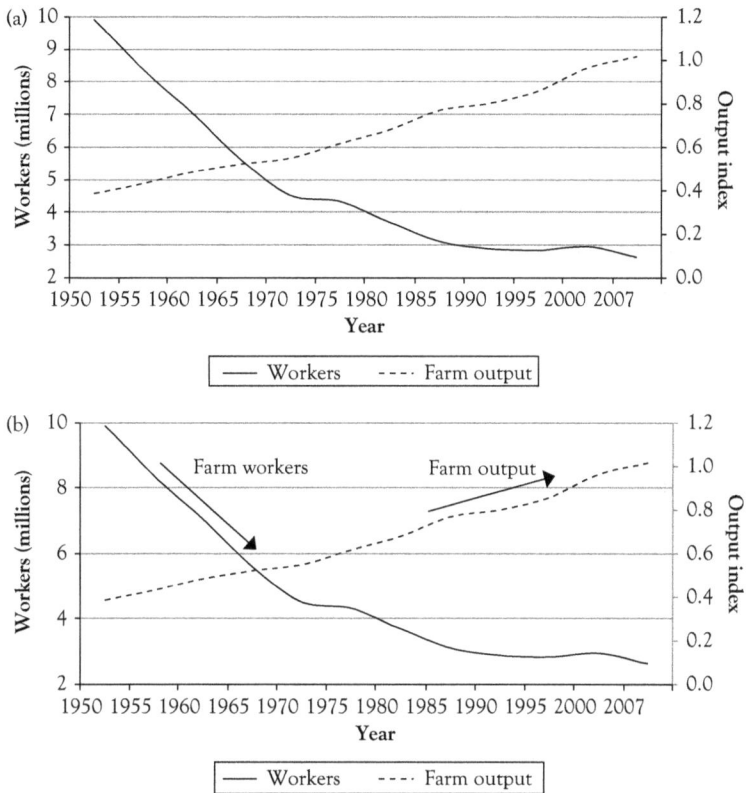

Figure 6.3 *Total farm output and workers*
Source: U.S. Department of agriculture (2013; n.d.).

Gains in productivity have been a driving force for growth in U.S. agriculture. The effects of these changes over the second half of the 20th century were dramatic: between 1950 and 2000, the average amount of milk produced per cow increased from 5,314 pounds to 18,201 pounds per year, the average yield of corn rose from 39 bushels to 153 bushels per acre, and each farmer in 2000 produced on average 12 times as much farm output per hour worked as a farmer did in 1950 (Fuglie, MacDonald, and Ball 2007).

Figures 6.3a and 6.3b illustrate the dramatic simultaneous drop in employment and the sustained growth of output.

For farming, not only did employment drop, but the number of farms dropped significantly. Figure 6.4 shows the simultaneous drop in farms and in employment. As shown in Figure 5.2, however, as the number of farms dropped, the average size of the farm increased but the land in farming remained fairly constant. Thus increased output was achieved with fewer workers, on larger farms, but with the overall acreage under farming remaining relatively constant.

Figure 6.5 presents data from the USDA showing indexes for total farm output and for labor input, fertilizer and lime input, pesticides input, and capital input. The most striking aspect of this figure is the

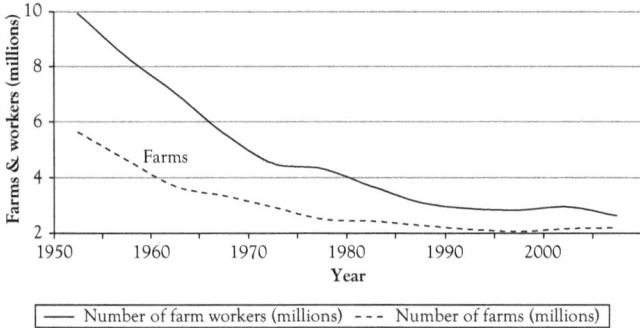

Figure 6.4 Total U.S. farms and farm workers
Source: U.S. Department of agriculture (n.d.).

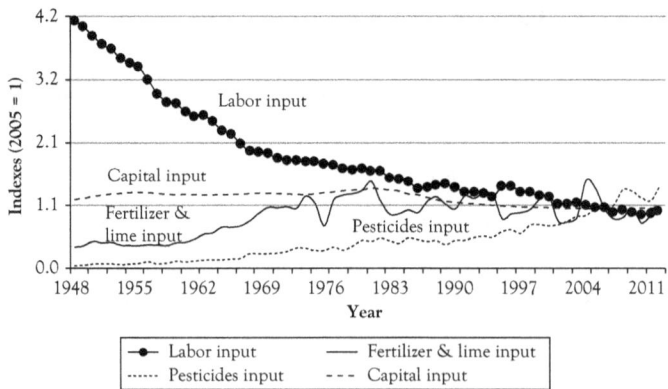

Figure 6.5 Indexes for total farm key inputs
Source: Fuglie, MacDonald, and Ball (2007).

Table 6.1 Farm output and input indexes. Total change and average annual cumulative growth rate

	Index of Total Farm Output	Index of Labor Input	Index of Fertilizer & Lime Input	Index of Pesticides Input	Index of Capital Input
1948–2012	155.90%	−78.14%	190.43%	3021.54%	−12.66%
AACGR	1.50%	−2.38%	1.71%	5.61%	−0.21%

Source: U.S. Department of argiculture (2013).

decline of the labor input index from four to one, reflecting the drop in the number of workers that is seen in Figures 6.2 and 6.3. Capital input (land, equipment, buildings, and so forth) declined after 1980 but has been more or less stable for the past 20 years.

Table 6.1 summarizes the total percentage and the average annual cumulative growth rate (AACGR) for the farm output and input indexes. As may be seen the dominant contributors to farm productivity growth have been the sizeable growth in the use of fertilizer and, more dramatically, pesticides with a percentage growth in that input index of over 3000 percent. With this growth in productivity arising from fertilizer and pesticides, farm output was able to increase at 1.5 percent per year even with labor input falling at 2.4 percent a year.

This is the agricultural model that has often been used as an analogy to manufacturing—that productivity gains in manufacturing from inputs such as information technology and robotics have enabled growing output with fewer workers. The decline in manufacturing employment is nothing to be concerned about because output is up. It must be noted, however, that the trends for employment and establishments in manufacturing appear very different from the trends in farming. Figures 6.6a and 6.6b present the manufacturing employment and the number of establishments since 1980. As may be seen, during the 80s and 90s, total employment was stable with a slight decline but the number of manufacturing establishments grew strongly. As may be seen, however, around 1999 and 2000, a sudden change occurred—both employment and establishments began a sharp decline that is quite different from any trends seen in agriculture. So is manufacturing truly OK? The validity of this question is now examined.

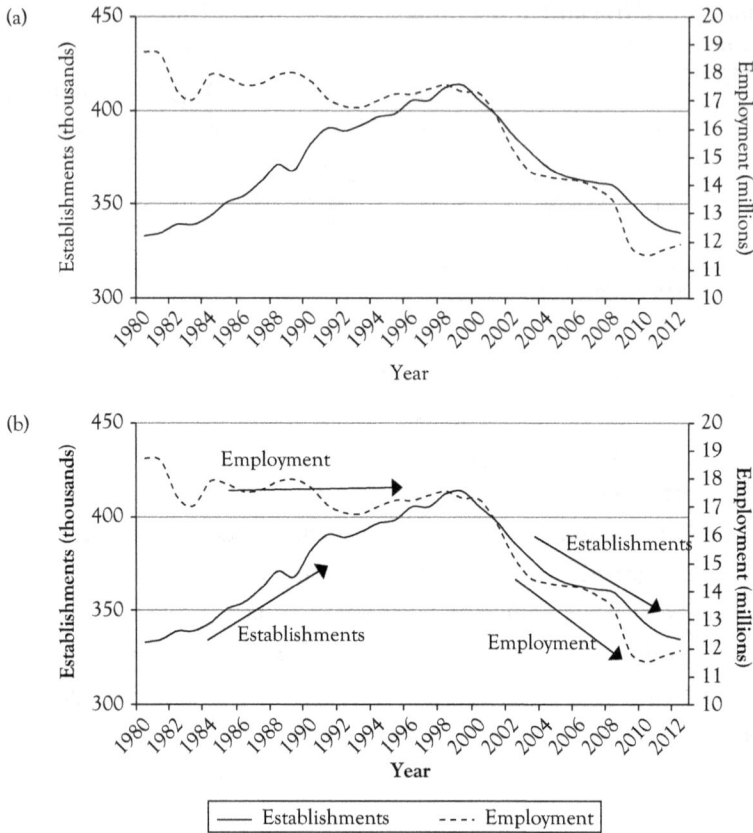

Figure 6.6 Manufacturing establishments and employment
Source: Bureau of Labor Statistics (2014a).

Measuring Output for Manufacturing Industries—It's Harder than It Looks

"We honestly don't have a clue about what's really going on in the U.S. economy. What's worse, we think we do" (Mandel 2012).

As shown in the previous section, output for the agriculture sector grew substantially in the face of falling employment. Is that also the case for manufacturing?

This apparently simple question, however, raises a very troubling issue—what measure should be used for manufacturing output and for the production of specific manufacturing industries? As noted by Atkinson (Atkinson et al. 2012), there is a confusing array of different measurements. Output in typical usage can refer to either gross output or value added. Gross output is essentially gross sales or receipts—basically gross revenues. Value added equals gross output minus intermediate inputs, that is, gross receipts from production less the value of goods and services used in the production. Nominal value, which is sometimes referred to a current value, is the raw figure that has not been adjusted for product prices or quality that change over time. Real value is raw figures adjusted for price and quality changes over time. Most assessments of manufacturing make use of the metric real value added as compiled by the Bureau of Economic Analysis (BEA). Another metric, however, is the IP index published by the Federal Reserve. Both real value added and the IP index attempt to measure real output adjusted for both price and quality changes which may include gains in technical performance over time. In this situation, both increases in the number of units shipped and improvements in product quality or technical performance cause the real output metrics to rise.

As noted in a report prepared by the Congressional Research Service "Hollowing Out" in U.S. Manufacturing: Analysis and Issues for Congress (Levinson 2013)

> Government statistical agencies address this problem by making highly technical adjustments when measuring certain prices. These adjustments can affect prominent economic indicators, such as gross domestic product (GDP) and labor productivity. The industries for which data are adjusted in this way, such as semiconductor manufacturing, are among the most vigorous in U.S. manufacturing, leading to questions about whether reported improvements in manufacturing represent real changes or are merely the result of statistical adjustments.

These concerns about the real health of manufacturing are reflected also in "Worse Than the Great Depression: What Experts Are Missing About American Manufacturing Decline" (Atkinson et al. 2012)

> Lamentably, the state of American manufacturing—and by extension the American economy—has been seriously misdiagnosed. In fact, the idea that "all is well" is faulty on two counts. First,

even when relying on official U.S. government data, it is clear that manufacturing output growth has lagged this decade, particularly in a number of key sectors. Second, and more importantly, it is increasingly clear that there are substantial upward biases in the U.S. government's official statistics and that real manufacturing output and productivity growth is significantly overstated. The most serious bias relates to the computers and electronics industry (NAICS 334)—its output is vastly overstated.

Unfortunately, these two official government statistics, real value added and the IP index, can yield quite divergent results as well as being overstated. Two manufacturing subsectors that have been plagued by these measurement issues are the computer and electronic products subsector (NAICS 334) and the petroleum and coal products subsector (NAICS 324). Tables 6.2a and 6.2b present the percentage change in the IP index, the real value added, and the gross output for the computer and electronic products subsector and the petroleum and coal products, respectively. Note again that both the IP index and the real value added attempt to measure in "real" terms, in a constant dollar sense, such that changes in price are removed from the change in market value. This output is then adjusted to include changes in the product's technical capabilities or quality. The gross output, however, is in current dollars, not deflated, so one might expect higher growth in output for this metric. That is not, however, what is seen here.

Table 6.2a *Comparison of output measures for computer and electronic products (NAICS = 334)*

Period	
1990–2011	Percentage Change for Industrial Production, FRB, 2007=1
	1601.23%
1990–2011	Percentage Change for Real Value Added, BEA, Billions of Chained (2005) dollars
	5140.58%
1990–2011	Percentage Change in Gross Output (Current Market Dollars)
	39.76%

Source: The Federal Reserve (2013).

Table 6.2b Comparison of output measures for petroleum and coal products (NAICS = 324)

Period	
1990–2011	Percentage Change for Industrial Production, FRB, 2007=1
	22.86%
1990–2011	Percentage Change for Real Value Added, BEA, Billions of Chained (2005) dollars
	171.10%
1990–2011	Percentage Change in Gross Output (Current Market Dollars)
	369.86%
1990–2011	U.S. Refinery & Blender Net Production of Crude Oil & Petroleum Products (Thousand Barrels)
	22.26%

Source: The Federal Reserve (2013).

For computers and electronic products, the IP index as estimated by the Federal Reserve increased by 1,601 percent over the 21 years between 1990 and 2011. The real value added of this industry was estimated by the BEA to have increased by a remarkable 5,141 percent. Both of these numbers of course incorporate the dramatic growth in technical capabilities over this period. Figure 6.7, Moore's Law, shows the exponential growth in the number of transistors on a microprocessor. In 1990, the count was approximately one million. By 2011 and 2012, the number had grown to at least a billion, if not somewhat greater. This rapid growth in capability (speed) created most of the growth in value added and in the IP index. It should be noted that the gross output or gross sales of the U.S. manufactured computers (in current dollars!) only grew by 40 percent over this time period. For an industry that lost roughly 800,000 employees and had very slowly growing sales over this period, a growth in value added of over 5,000 percent seems highly misleading. Even though the IP index is up by 1,601 percent it seems to be a more reasonable figure to use but it also overstates the size and health of this industry.

For petroleum products, the comparison is not so dramatic. Value added is up by a robust 171 percent but the IP index increased by only 23 percent. This index growth closely aligns with the increase in the actual number of barrels of product produced. For these reasons, the Federal Reserve's IP index will be used as the metric for manufacturing output

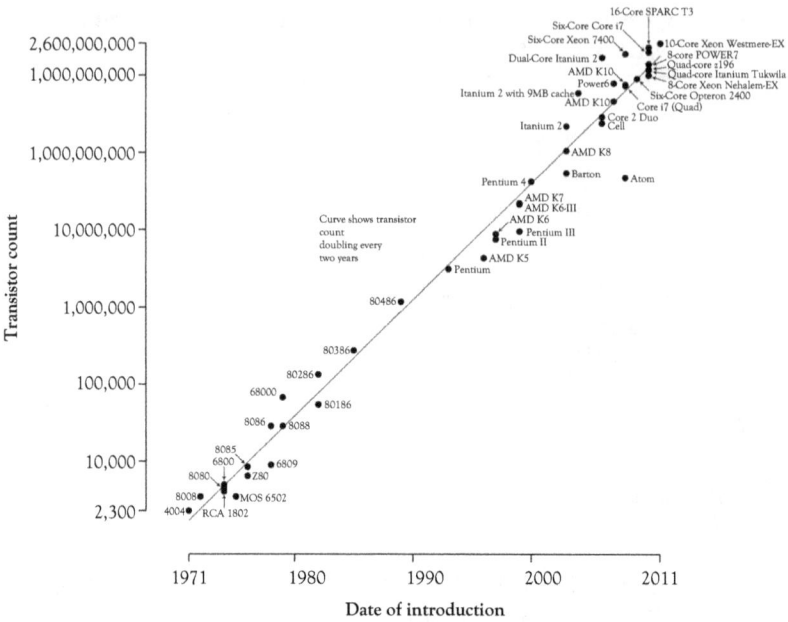

Figure 6.7 *Microprocessor transistor counts 1971–2011 and Moore's law*

over time. For additional discussion of the challenges of value added see "Why Value Added is Not My Favorite Metric" by Meade as well as the paper by Atkinson.

Industrial Production Index for Manufacturing Industries

It turns out in fact that the IP index for all of manufacturing, as shown in Figure 6.8, has grown substantially—even during the period from 2000 to 2007 when so many jobs and establishments were lost. The negative impacts of the recent recession are clearly evident but so is the rebound. Over the period 1972–2012, real GDP increased by 201.64 percent at a real annual cumulative growth of 2.80 percent per year. From 1972 to 2013, the manufacturing index increased by 171 percent in real terms or a cumulative average rate of 2.46 percent. Although growing somewhat less than the overall economy, the figures for manufacturing appear to be respectful growth numbers. So is manufacturing really OK?

At the next level of detail, however, Figure 6.9 shows a clear difference between the industrial output for the manufacturing of durables and
nondurables. On one hand, nondurable manufacturing has a much slower
growth in output than the manufacturing of durables and the sector has
a much weaker recovery. On the other hand, manufacturing of durables
shows very strong growth in production from 1990 to 2007 and then a
much stronger rebound from the recession. Over the 41-year period from
1972–2013, the index for IP of nondurable manufacturing only increased

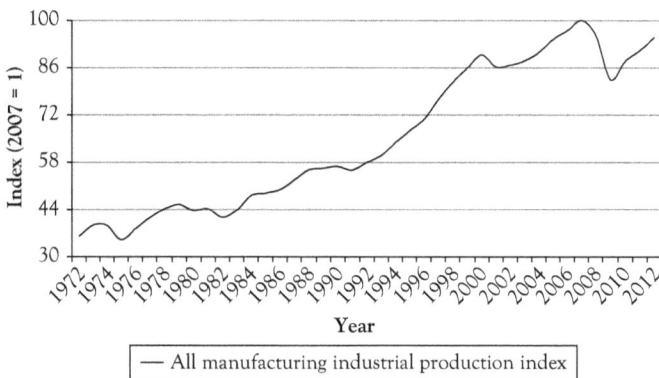

Figure 6.8 Industrial production index for all manufacturing
Source: The Federal Reserve (2013).

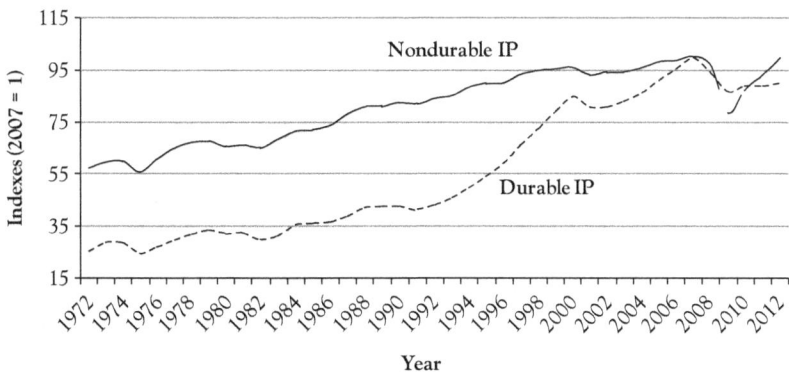

Figure 6.9 Industrial production indexes for durable and nondurable
manufacturing
Source: The Federal Reserve (2013).

by 59 percent or an average annual rate of just 1.14 percent. The durable manufacturing index increased a remarkable 307 percent or 3.49 percent per year, significantly higher than the overall economy. This appears to be a very healthy growth rate. As with both employment and establishments in earlier chapters, however, it is important to examine IP for the durable and nondurable industries in order to understand the higher level metrics.

Industrial Production for Subsectors of Nondurable Manufacturing

It has been seen in earlier chapters that the nondurable subsectors of apparel and leather and allied products experienced massive losses of jobs and manufacturing plants. The agricultural model might surmise that these losses were the result of productivity gains. A key question here is whether output has gone up as in agriculture. Figure 6.10 shows that the production of leather goods and allied products in the United States (for example, shoes) has been, in fact, in free fall since the early seventies, with industry output declining by 76 percent. As may be seen in Table 2.1, 103,800 jobs have been lost in the production of leather goods since 1990. The apparel production trend in Figure 6.10 has a very different pattern—it is relatively stable until the mid-nineties, then begins

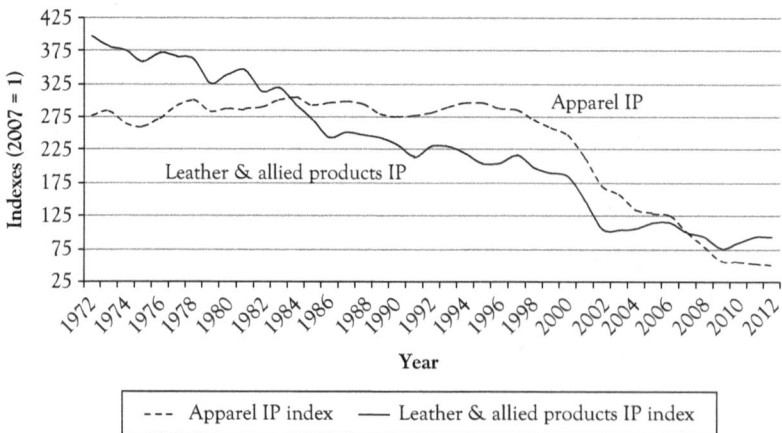

Figure 6.10 Industrial production indexes for apparel and leather manufacturing

Source: The Federal Reserve (2013).

to decline and after 2000 starts a precipitous decline. Since its peak in 1996, apparel production has declined by 82.9 percent. As presented in Table 2.1, since 1990, apparel production has lost 754,700 jobs, a decline of 83.6 percent, a percentage about equal to the drop in production.

Figure 6.11 presents indexes for apparel output and employment with 1990 = 1. As may be seen, output increases somewhat in the early 90s while employment is already falling. This would indeed be a period of rising productivity as management strives to maintain competitiveness with layoffs and closings of least efficient plants. By 1995, however, output begins to decline and employment then begins to plummet. Thus, in this case, employment dropped following the declining demand and output rather than from increased productivity. These manufacturing industries are certainly not in good shape and, in fact, have essentially disappeared from the U.S., and employment has dropped not because of productivity gains but because of reduced output.

With the loss of the apparel manufacturing, a major supplier subsector was inevitably impacted. Figure 6.12 presents the IP indexes for fibers, yarn, and thread; fabric mills; and textile and fabric finishing mills. These industries actually showed some modest growth in production from the mid-eighties to the mid-nineties, then the decline begins—following apparel down the steep slope. Following its peak in 1998, the production

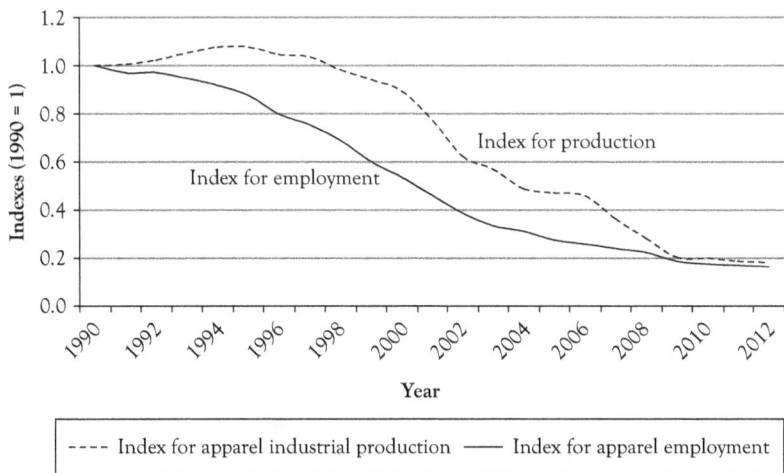

Figure 6.11 Indexes for apparel production and employment
Source: The Federal Reserve (2013).

of fiber, yarn, and thread had declined by 56.6 percent by 2012. Production at fabric mills peaked in 1995 and by 2012 had declined by 50.2 percent. Textile and fabric finishing mills and fabric coating production peaked in 1994 and by 2012 had declined by 58.9 percent. Between 1990 and 2012 these industries had shed 373,800 jobs as presented in Table 2.1.

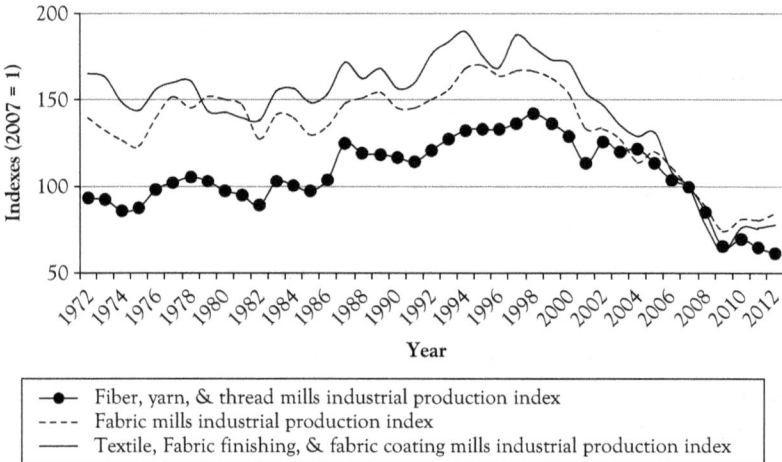

Figure 6.12 Industrial production indexes for textiles and fabrics
Source: The Federal Reserve (2013).

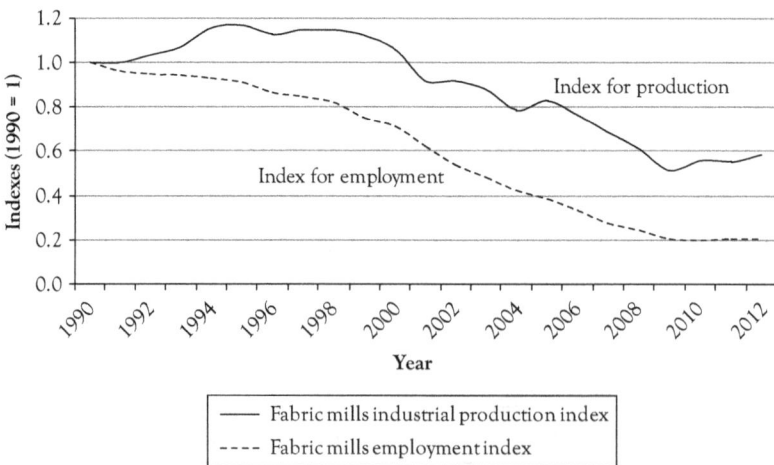

Figure 6.13 Indexes for fabric mills production and employment
Source: The Federal Reserve (2013).

Figure 6.13 presents indexes for fabric mills production and employment. As with apparel, there is a period in the early nineties with rising output and declining employment, indicating higher productivity gained either through layoffs and closing of least efficient plants. After 1998, both output and employment began a long steep decline.

Other nondurable subsectors that have experienced declining output and demand are the paper production and printing. As Figure 6.14 shows, the paper and paper products production index has fallen by 24 percent from its peak in 1999, and the printing production index has fallen by 29 percent from its peak in 2000. This decline in paper manufacturing and printing production most likely reflects the impacts of the Internet on printed material such as newspapers and magazines as well as the development of high quality laser and inkjet printers that can be driven by personal computers at the office or home. As shown in Tables 2.1 and 3.1, between 1990 and 2012, paper and paper products production shed 268,200 jobs and printing and related support activities lost 346,400 jobs, losses of 41 percent and 43 percent, respectively. Over that time period, 18,181 printing and related support establishments were lost, a 38 percent drop in establishments.

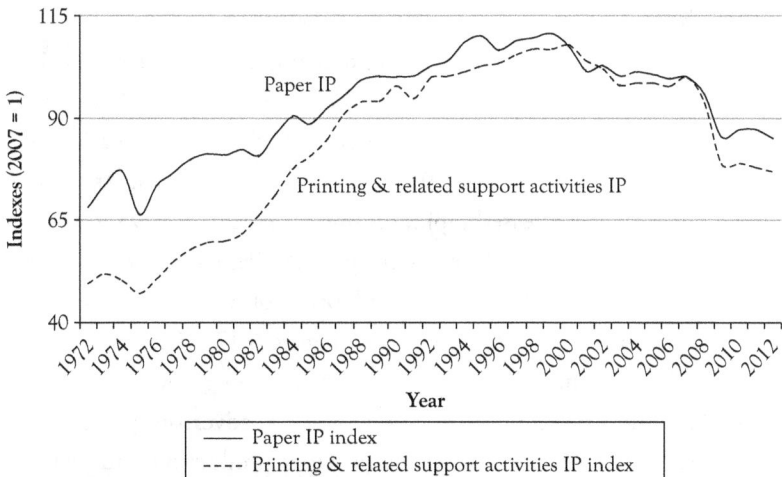

Figure 6.14 Industrial production indexes for paper and printing and related support activities

Source: The Federal Reserve (2013).

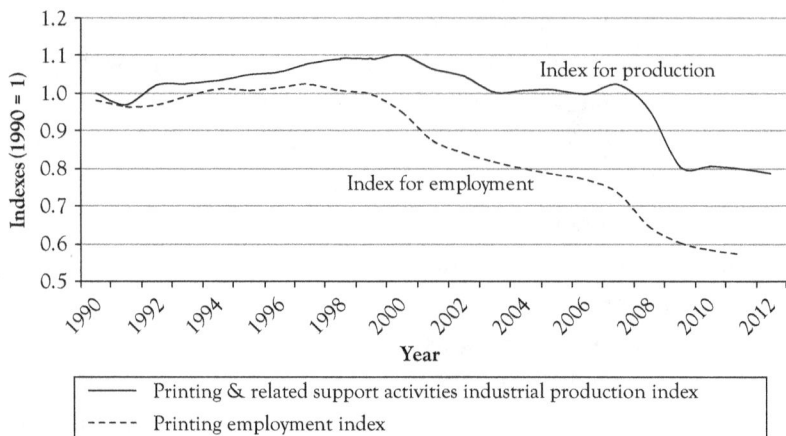

Figure 6.15 Indexes for printing production and employment
Source: The Federal Reserve (2013).

Figure 6.15 presents indexes of output and employment for printing and related support activities. As may be seen, prior to the year 2000, production and employment were somewhat increasing with output being up slightly more, reflecting greater productivity. However, by 2000, output began a decline and then experienced a large reduction in the Great Recession and has seen little recovery. Employment in the printing industry has been in decline since 2000. There have been some productivity gains over the period but declining demand and output have driven down employment.

Figure 6.16 presents the percentage of the U.S. adult population who use the Internet from June of 1995 through May of 2013. As may be seen, the usage rate increased rapidly from 14 percent in 1995 to 50 percent in 2000 and then to 72 percent by 2005. Perhaps more impactful on paper and printing was the rapid adoption of broadband that began around the year 2001 as may be seen in Figure 6.17. High speed access made the Internet much more effective in providing news, exchanging documents, and providing information as well as advertising. There have clearly been some productivity gains in paper production and printing but the dominant cause of employment loss has not been productivity but declining demand and output. Again this case represents an industry that is certainly not the agricultural model or evolution.

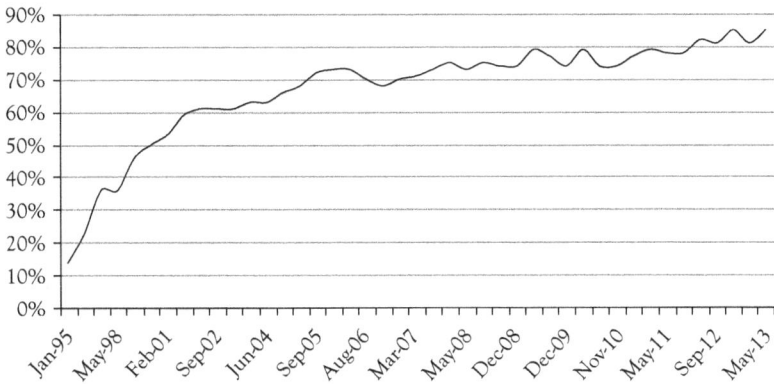

Figure 6.16 Internet adoption (% of American adults who use the Internet, over time)

Source: Pew Research Center (n.d.).

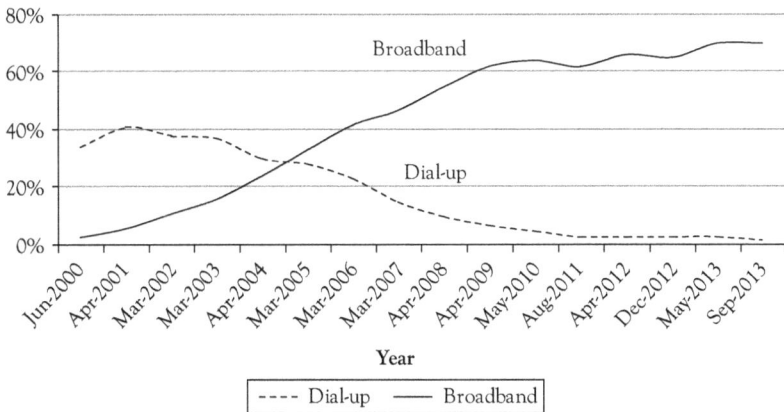

Figure 6.17 Home broadbrand versus dial-up (2000–2013). Percentage of American adults 18 years and older who access the Internet

Source: Pew Research Center (n.d.).

Figure 6.18 presents the production indexes for food, beverage, and tobacco products. The most eye-catching trend in this figure is the dramatic drop in production of tobacco products, a drop of 49.7 percent. This clearly arises from the declining number of smokers in the United States. As shown in Figure 6.19, the percentage of adults that smoke dropped from over 40 percent of adults to less than 20 percent between

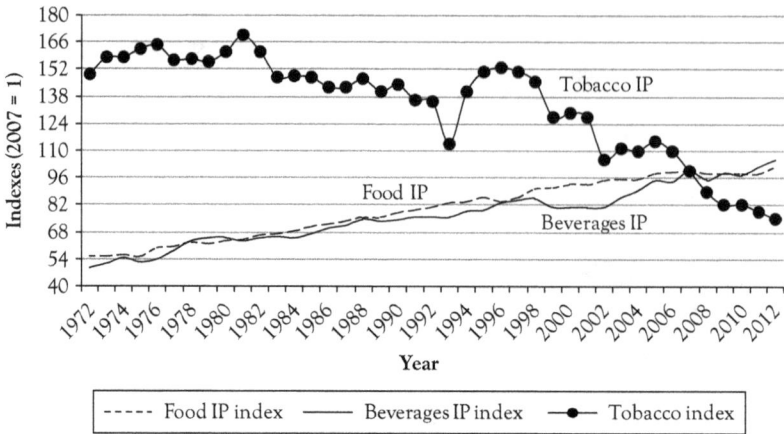

Figure 6.18 Industrial production indexes for food, beverage, and tobacco production

Source: The Federal Reserve (2013).

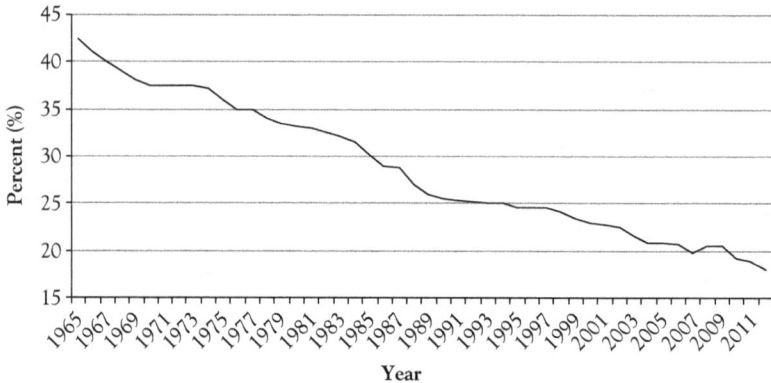

Figure 6.19 Trends in adult smoking (1965–2012)

Note: Percentage of adults who are current cigarette smokers (National Health Interview Survey 1965–2012)

Source: Centers for Disease and Prevention (n.d.).

1965 and 2011. This decline in smokers created an attendant decline in demand, and consequently, output by the industry. As a result, employment in the tobacco industry dropped by 68 percent. In contrast, the food IP index increased by 83.5 percent and beverage production by 112.1 percent. With greatly reduced output and plunging employment, the tobacco product manufacturing industry is certainly not following the agricultural model.

Figure 6.20 presents the historical indexes for IP for petroleum, chemicals, and plastics and rubber products. These subsectors showed growth in IP in the period through the seventies, eighties, and nineties. Petroleum products and chemicals both peaked in 2007 before the Great Recession and have shown modest recovery since then. Over the 1972–2007 period, petroleum products production increased by 52.6 percent and chemicals increased 143.1 percent. Plastic and rubber products production peaked in 2006 and had increased by 196 percent since 1972. From their peak before the recession, petroleum product production was 4.35 percent down in 2012, chemicals were down 13.6 percent. Plastics and rubber were still 15.7 percent below their peak in 2012.

Figures 6.21, 6.22, and 6.23 present the three metrics for petroleum and coal products—the IP index, the real value added, and the net production. Table 6.3 summarizes the changes over time for these three metrics. As may be seen, the percentage change in value added is significantly greater than both the IP index and the barrels of product produced, which are both close in value. This is another reason why this analysis tends to focus on the Federal Reserve IP index rather than the BEA value added.

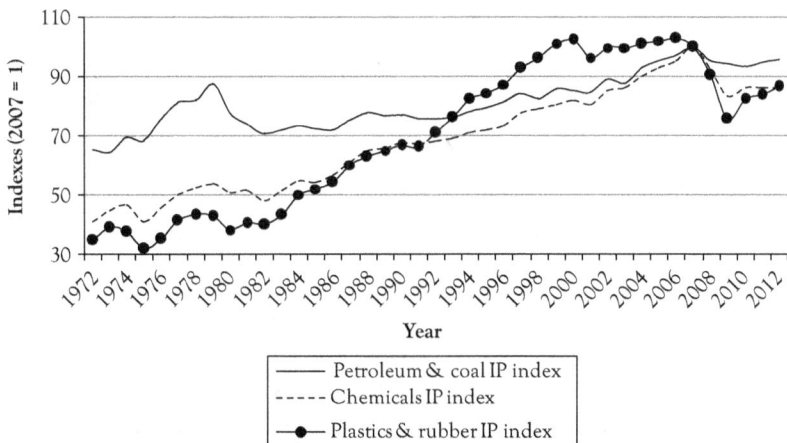

Figure 6.20 Industrial production indexes for petroleum and coal, chemicals, and plastics and rubber products

Source: The Federal Reserve (2013).

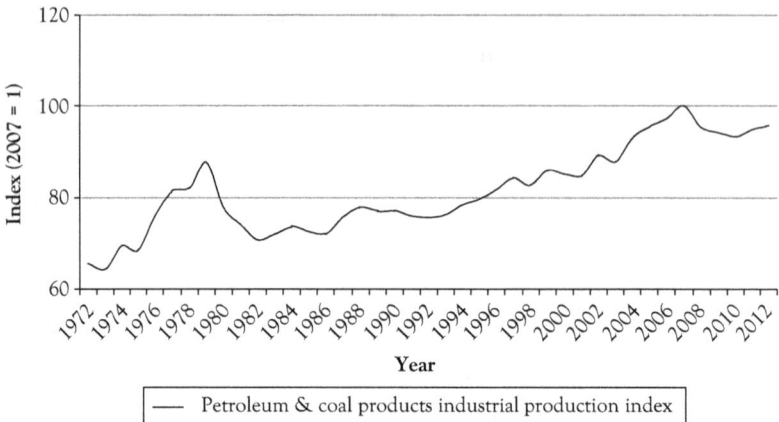

Figure 6.21 Industrial production index for petroleum and coal products

Source: The Federal Reserve (2013).

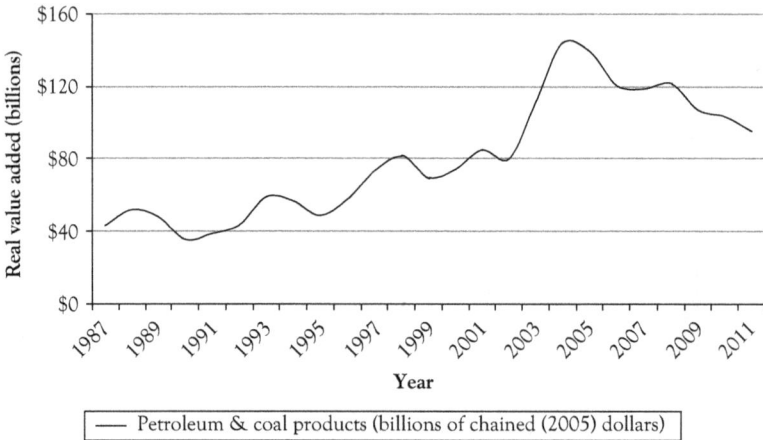

Figure 6.22 Real value added for petroleum and coal products

Source: Bureau of Economic Analysis (n.d.).

Table 6.4 presents an overview of the changes in IP and employment for the nondurable manufacturing industry. In brief, only the beverage industry showed an increase in employment and that was for only 5,100 jobs. The other 11 subsectors and industries lost a total of two and a half million jobs between 1990 and 2012. There is no simple, single reason for

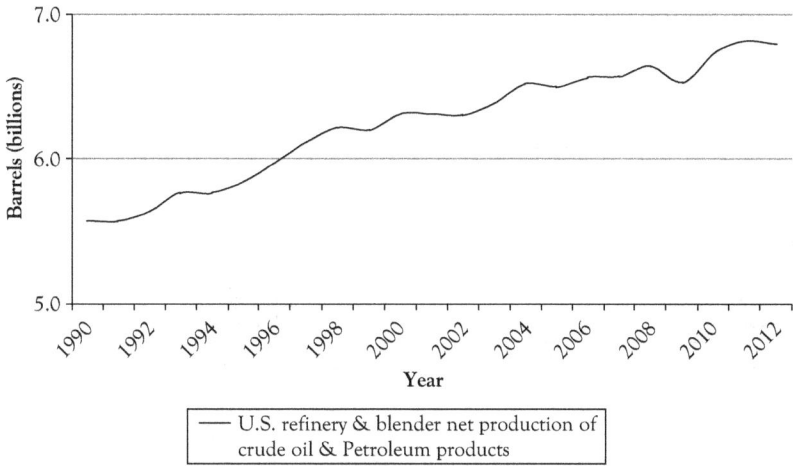

Figure 6.23 U.S. Refinery and blender net production of crude oil and petroleum products

Source: Energy Information Administration (2013).

Table 6.3 Metrics for manufacturing of pertoleum and coal products

	Total Percentage Change 1987–2011	Average Annual Growth Rate 1987–2011
Industrial Production Index	26.68%	0.99%
Value Added	121.02%	3.36%
Barrels of Product Produced	27.66%	1.02%

Source: Bureau of Economic Analysis (n.d.).

these losses. But one thing is clear; these jobs were not lost primarily due to productivity growth, that is, by following the agricultural model. Dominant causes in the cases of tobacco products, paper and printing, and apparel and textiles were the rapid decline in smoking, the rapid adoption of broadband Internet access, and the importation of clothing, each cause creating a reduction in output of U.S. manufacturing. In other industries, productivity surely had a key role. As may be seen in Table 6.4, petroleum products, chemical, and plastics and rubber products had significant increases in their production indexes that were nevertheless accompanied by significant reductions in employment. These types of plants have become increasingly automated enabling doing more with less.

Table 6.4 Change in nondurable industrial production and employment

	% Change in Industrial Production Index			Percentage Change in the Number of Jobs	Change in Number of Jobs
	1972–2007	2007–2012	1990–2012	1990–2012	1990–2012
Food (NAICS = 311)	79.8%	2.1%	30.8%	–2.6%	–38,600
Beverage (NAICS = 3121)	100.0%	6.1%	42.2%	3.0%	5,100
Tobacco (NAICS = 3122)	–32.9%	–25.0%	–48.0%	–68.0%	–30,600
Textile Mills (NAICS = 313)	–26.0%	–21.7%	–44.9%	–76.0%	–373,800
Textile Product Mills (NAICS = 314)	23.0%	–28.5%	–35.7%	–50.5%	–119,000
Apparel (NAICS = 315)	–63.8%	–50.8%	–82.1%	–83.6%	–754,700
Leather & Allied Products (NAICS = 316)	–74.9%	–6.6%	–59.9%	–77.9%	–103,800
Paper (NAICS = 322)	46.7%	–14.8%	–14.9%	–41.4%	–268,200
Printing & Related Support Activities (NAICS = 323)	102.4%	–23.1%	–21.4%	–42.8%	–346,400
Petroleum & Coal Products (NAICS = 324)	52.6%	–4.4%	24.0%	–25.9%	–39,600
Chemicals (NAICS = 325)	143.1%	–13.6%	28.0%	–24.3%	–252,100
Plastics & Rubber Products (NAICS = 326)	187.8%	–13.3%	29.7%	–21.8%	–179,600
TOTAL JOBS LOST IN NONDURABLE MANUFACTURING –2.501,300					

Source: The Federal Reserve (2013).

Industrial Production for Subsectors of Durable Manufacturing

Over the period 1972–2007, just before the impacts of the recent recession, the IP index for durable manufacturing had increased by a remarkable 293 percent compared to only 75 percent for nondurables. Following the recession, durables were down from 2007 levels by only 0.13 percent. In contrast, nondurables were down by 10.5 percent in 2012 from 2007. These trends may be seen in Figure 6.9. An important question is what accounts for the strong growth in the manufacturing of durable goods? Durable manufacturing has the following subsectors: wood products; nonmetallic mineral products; primary metals; fabricated metal products; machinery; computer and electronic products; electrical equipment, appliances, and components; motor vehicles, bodies and trailers, and parts; other transportation equipment; furniture and related products; and miscellaneous manufacturing. Once again, major differences emerge at the sector level.

Figure 6.24 presents the historical trend line for the IP index for wood products. This index appears to have three distinct periods: 1972–1990, from 1990 to the peak of this index in 2006 with the housing boom, and then the period from 2006–2012. Between 1972 and 1990, the wood product production index increased by only 14.1 percent or at

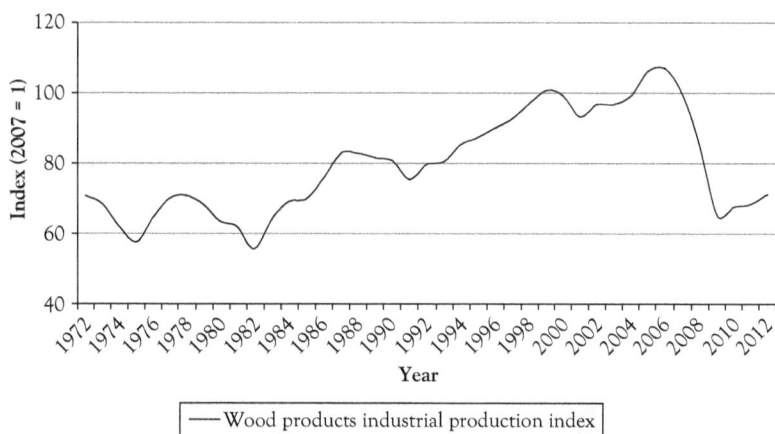

Figure 6.24 Industrial production index for wood products
Source: The Federal Reserve (2013).

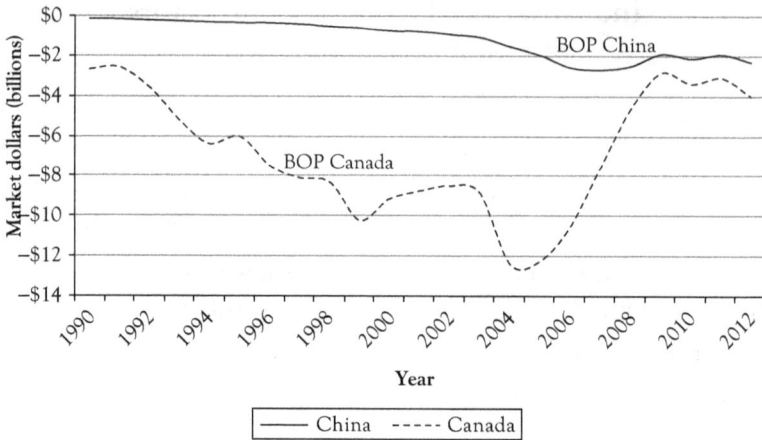

Figure 6.25 Balance of payments for wood products—China and Canada

Source: ITA Trade with Selected Market (2014)..

the very slow annual rate of 0.74 percent per year on average and then between 1990 and 2006, the wood products production index increased by 32.5 percent or an annual average rate of 1.77 percent. With the housing bust and the financial crisis, this index then fell by 33.3 percent between 2006 and 2012 and is recovering slowly. It is unlikely that many Americans think in terms of the U.S. importing wood products such as lumber, plywood, and veneer but it is significant. Although trade and the balance of payments (BOP) will be addressed in Chapter 7, it is interesting to note that from 1990–2005, the U.S. BOP for wood products increased from a negative $1.2 billion to a negative $18.9 billion, an increase of 1,427 percent over just 15 years. Figure 6.25 presents the BOP for Canada and China, the two countries with the largest negative trade balances in wood products. Imports from Canada and the growing negative trade balance seem to increase substantially with the North American Free Trade Agreement going into effect on January 1, 1994. These trade issues will be addressed in more detail in Chapter 7, but it again raises questions about the loss of jobs in manufacturing being related primarily to productivity gains.

Figure 6.26 presents time series of indexes for wood product production and employment with 1990 = 1. As may be seen, after a slight dip in the early nineties, both production and employment increased steadily until

1999. Since the production index increased more than that of employment, some growth of productivity occurred. After 2005, however, both output and employment began a sharp decline. Loss of employment here was due to dramatically reduced demand and output—not productivity.

Figure 6.27 presents the IP indexes for nonmetallic mineral products (clay, glass and glass products, cement and concrete products, and

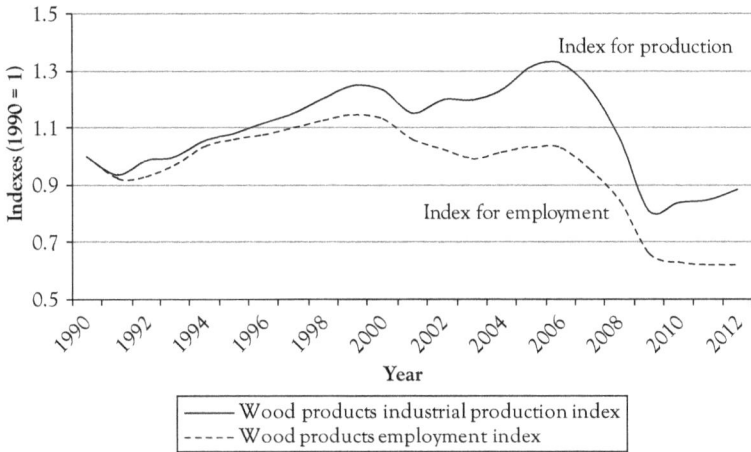

Figure 6.26 Index for wood products production and employment
Source: The Federal Reserve (2013).

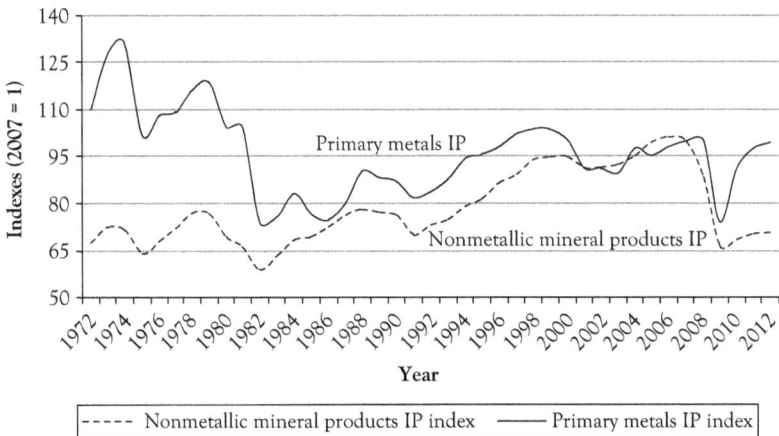

Figure 6.27 Industrial production indexes for nonmetallic mineral products and primary metals
Source: The Federal Reserve (2013).

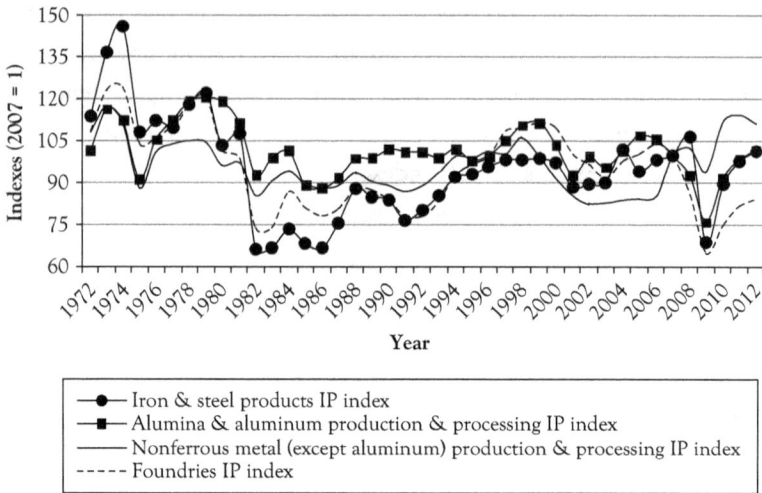

Figure 6.28 Industrial production indexes for primary metal industries

Source: The Federal Reserve (2013).

lime and gypsum) and for primary metals. The striking aspect of this figure is the lack of any real growth over a 40-year period, especially for primary metals. From 1972–2012, the real U.S. GDP increased by 201.6 percent—primary metals production, on the other hand, *declined* by roughly 10 percent over this period. This is very different from the agricultural model where output increased strongly over this period.

Figure 6.28 presents production indexes for the industries in primary metals. It is striking that the time lines are so similar for these four industries showing decline and stagnation, be it iron, aluminum, nonferrous metals, or foundries. This is a distressing figure. Table 6.5 presents the change in the IP indexes for these industries within primary metals. All four industries experienced a drop in production from 1972–1986. The period from 1987–1999 had modest growth then from 2000–2007 was stagnation in production. Foundries had the largest decline in production, and as shown in Table 3.2, over 30 percent of foundries closed between 1990 and 2012.

Figure 6.29 presents indexes for primary metal production and employment with 1990 = 1. As may be seen, after a slight dip in the early nineties, production increased until 1998. During that period

Table 6.5 Change in industrial production indexes for primary metals

Time Period	Iron & Steel Products (NAICS = 3311,2)	Alumina & Aluminum Production & Processing (NAICS = 3313)	Nonferrous Metal (Except Aluminum) Production & Processing (NAICS = 3314)	Foundries (NAICS = 3315)
1972–1986	–41.73%	–13.17%	–18.51%	–28.21%
1987–1999	30.32%	21.84%	12.13%	38.68%
2000–2007	2.80%	–3.44%	8.99%	–8.97%
2007–2012	1.09%	1.75%	11.37%	–15.57%
1972–2012	–11.33%	0.27%	3.06%	–22.37%

Source: The Federal Reserve (2013).

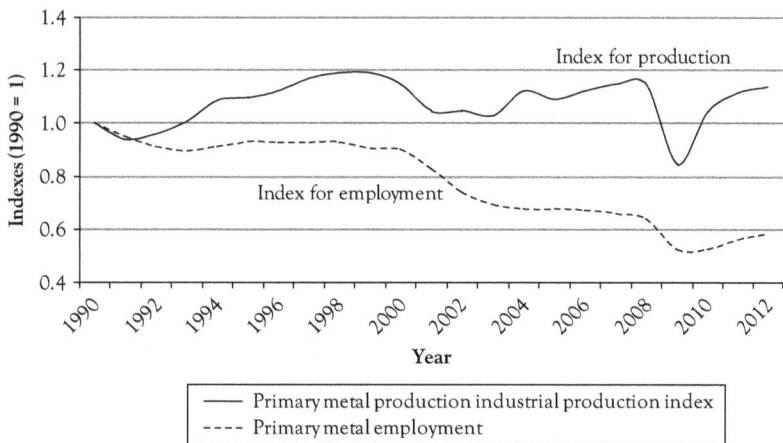

Figure 6.29 Indexes for primary metals production and employment
Source: The Federal Reserve (2013).

employment was lower, indicating some gains in productivity. After 1998, production fell and then was largely stagnant. Employment fell sharply after 2000. As shown in Figure 6.30, the BOP for primary metals was roughly negative $5 billion in 1993, grew to a negative $20 billion by 2000, and then grew to negative $50 billion by 2006–2007 before the recession. It appears that growth in demand is being satisfied by imports and that U.S. establishments have maintained stagnant levels of production with fewer and fewer employees.

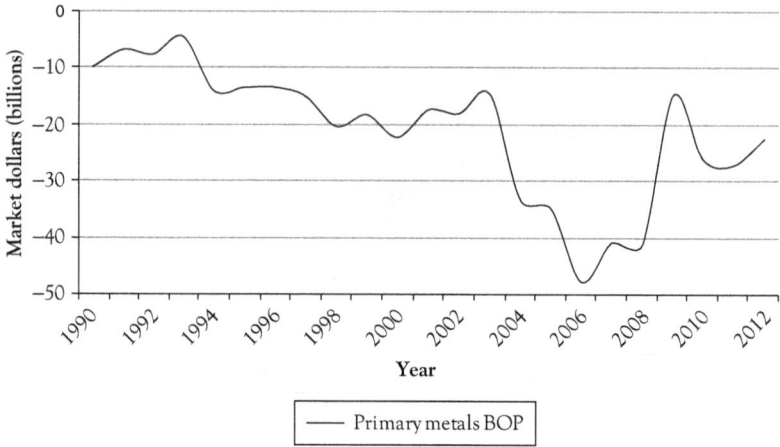

Figure 6.30 Balance of payments for primary metals
Source: ITA Trade with Selected Market (2014).

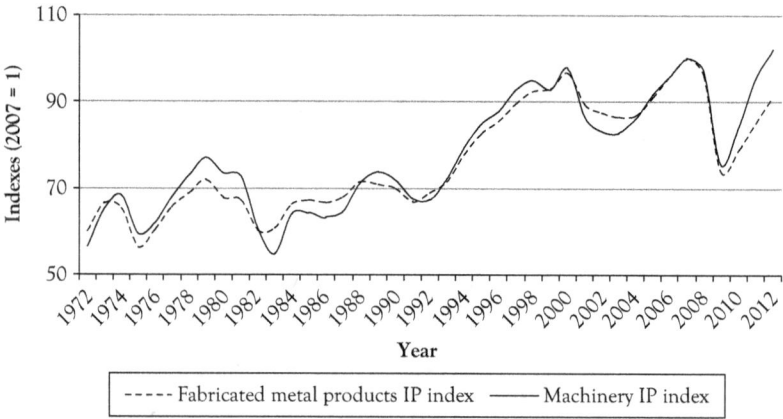

**Figure 6.31 Industrial production indexes for fabricated metal
and machinery products**
Source: The Federal Reserve (2013).

Figure 6.31 presents the IP indexes for fabricated metal products and
machinery. The striking thing about this figure is that, although there are
some ups and downs, once again there is stagnation after the year 2000.
This was also seen in the primary metals.

The stagnation is seen more clearly in Figures 6.32a, 6.32b, and
6.33. These figures present the Federal Reserve's data on IP and

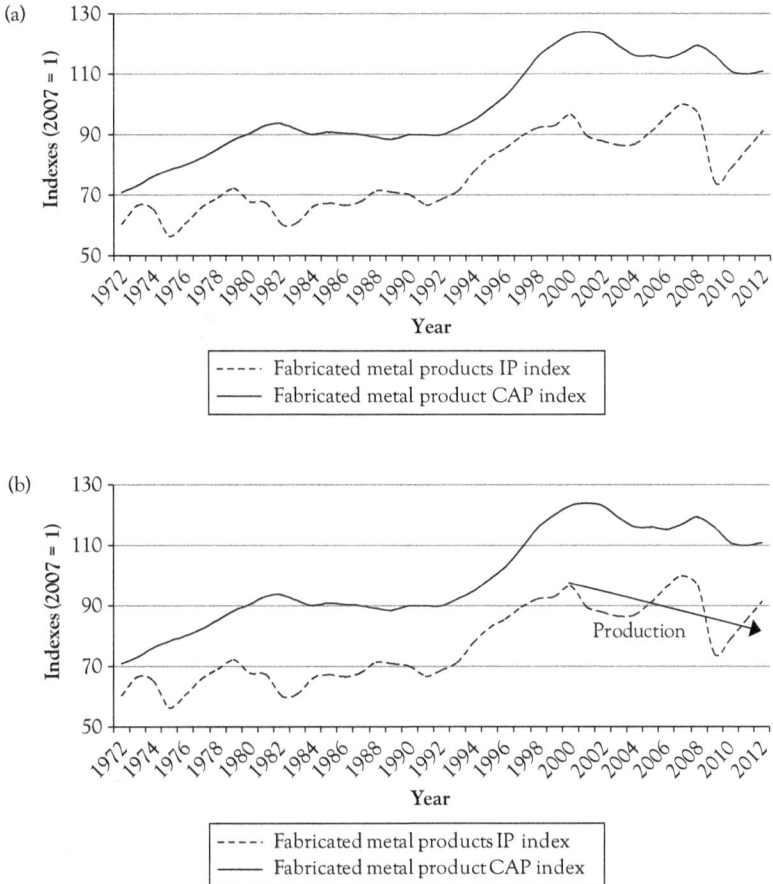

Figure 6.32 Industrial production and capacity indexes for fabricated products

Source: The Federal Reserve (2013).

industrial capacity for fabricated metal products and machinery. As may be seen, industrial capacity grew from 1972 until it peaked in 2000. Table 6.6 summarizes the change over time for these two industries and quantifies the slow growth rates. Capacity has been stagnant for machinery and declining for fabricated metal products. In other words, even in the face of a growing economy, these industries stopped growing. As may be seen, there was a period of stagnation in the 1980s followed by growth in the 1990s. The question now is whether growth

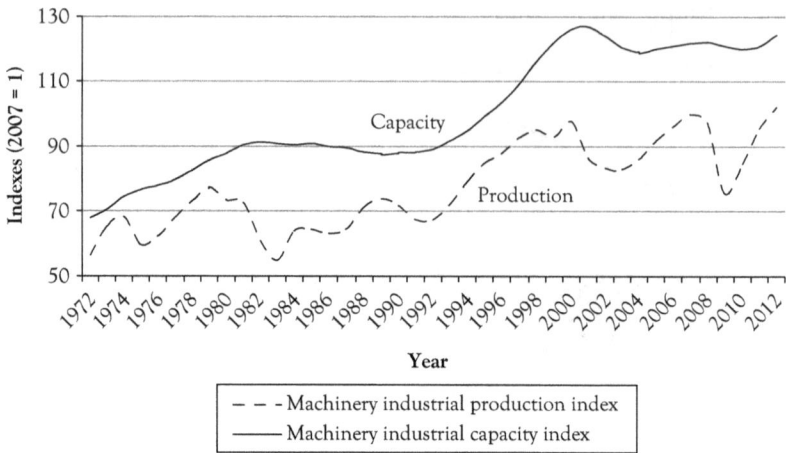

Figure 6.33 Industrial production and capacity indexes for machinery
Source: The Federal Reserve (2013).

Table 6.6 Change in industrial production indexes for fabricated metal products and machinery

	Fabricated Metal Products IP	Machinery IP
1972–2012	51.74%	81.20%
AACGR	1.05%	1.50%

Source: The Federal Reserve (2013).

is possible today or whether increased demand will be met through imports. These issues will be addressed in Chapter 7.

Figure 6.34 presents IP indexes for household and institutional furniture and for office furniture. As may be seen, production in these industries showed variable, but steady, growth from 1972 until, once again, a plateau is hit in the year 2000. Between 2000 and 2007, household furniture increased modestly but office furniture production declined. Figure 6.35 shows that furniture production capacity fell after 2000 with major reductions during the recession. This corresponds to the substantial decline in the number of furniture manufacturing establishments in Tables 5.10a and 5.11a. Overall, furniture manufacturing lost roughly 30 percent of establishments. Figure 6.36 presents indexes for furniture IP, capacity, and employment with 1990 = 1. As may be

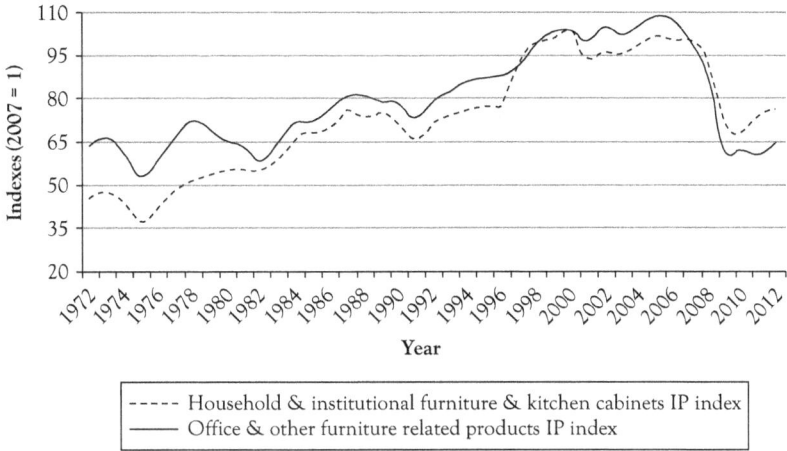

Figure 6.34 Industrial production indexes for household and institutional furniture and kitchen cabinets, and office and other furniture

Source: The Federal Reserve (2013).

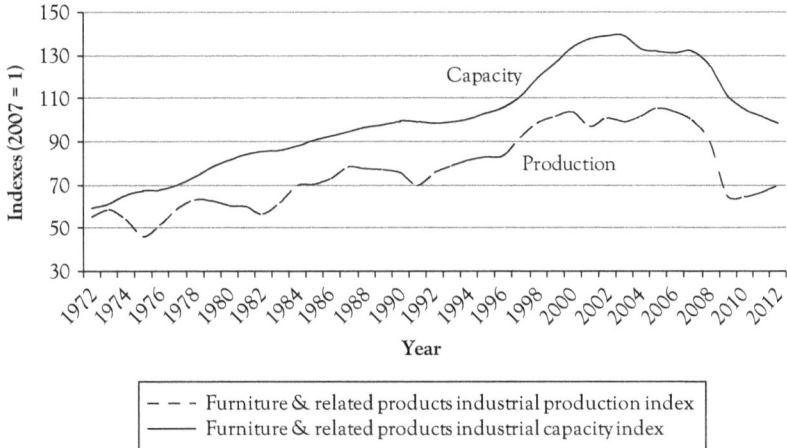

Figure 6.35 Industrial production and capacity indexes for furniture and related products

Source: The Federal Reserve (2013).

seen, from 1990 to 2000, production increased more than employment, indicating some gains in productivity. After the year 2000, however, production and capacity remained stagnant but employment began to

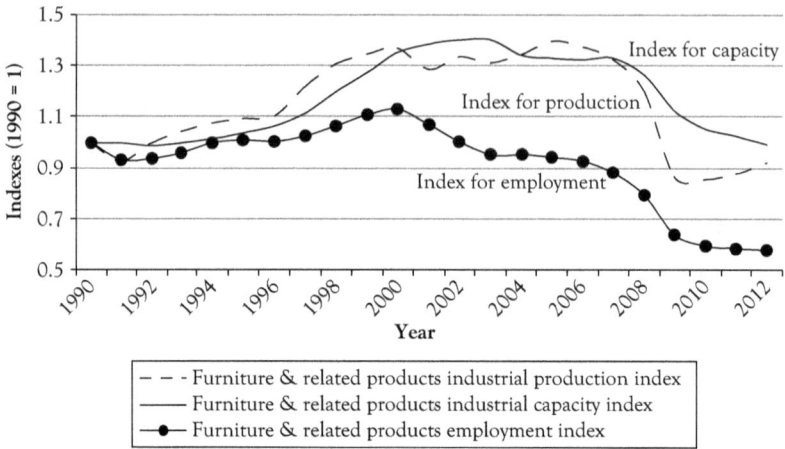

Figure 6.36 Indexes for furniture and related products, production, capacity, and employment

Source: The Federal Reserve (2013).

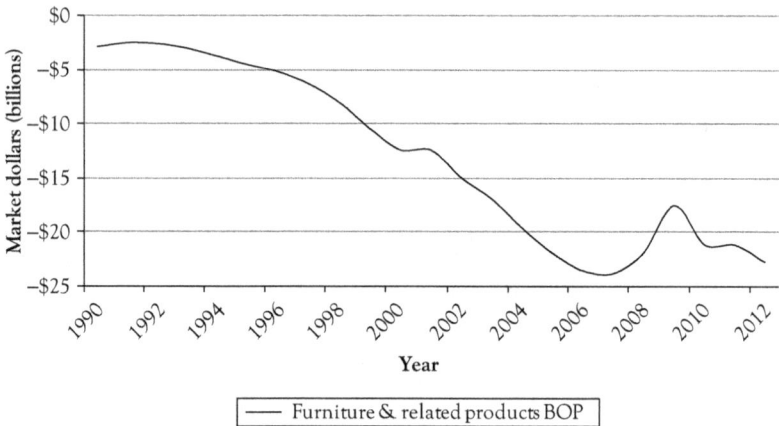

Figure 6.37 Balance of payments for furniture and related products

Source: ITA Trade with Selected Market (2014).

fall, reflecting layoffs and closing of the least-efficient plants. This was a time of soaring imports of furniture as seen in Figure 6.37, which presents the BOP for furniture. In 1995, the BOP for furniture was roughly a negative $4.5 billion, by 2000 that had grown to $12 billion

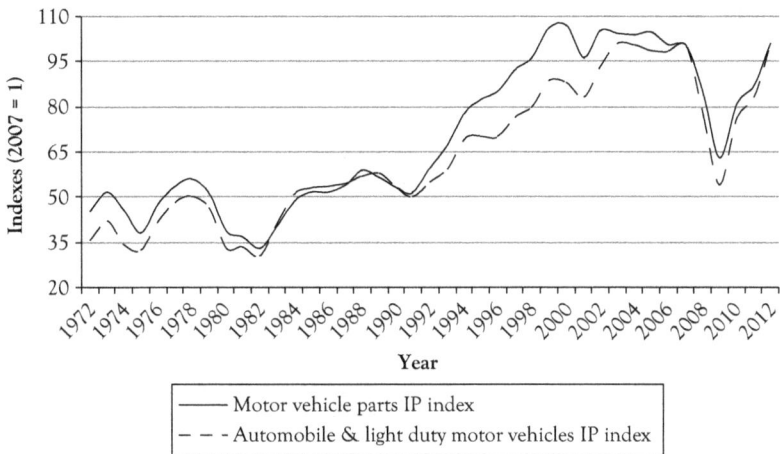

Figure 6.38 Industrial production indexes for motor vehicle parts, and automobile and light duty motor vehicle

Source: The Federal Reserve (2013).

and by 2007 it had soared to $24 billion. The U.S. furniture industry would now seem to be in a very difficult position. With production capacity lost, it is likely that future growth in demand will be largely met by imports.

Figure 6.38 presents IP indexes for automotive parts and for automobile and light duty motor vehicles. Once again, a pattern seen before emerges: slow and variable growth in the 1970s and 1980s, expansion in the 1990s, and then stagnation beginning around 2000. Again this pattern is seen more dramatically by examining industrial capacity. Figure 6.39 presents the IP index and the industrial capacity index for the manufacturing of motor vehicle parts. As may be seen, after decades of growth, industrial capacity for production of motor vehicle parts absolutely stagnates after 2000. What happens now with growth, will the U.S. automotive parts industry invest in the United States or will the growing demand be met by imports? Figure 6.40 presents the IP index and the industrial capacity index for the manufacturing of motor vehicles. In this figure, production capacity continues to grow after 2000 yet production tended to stagnate. It appears that there is now considerable excess capacity in this industry which will support growth in demand without excessive investment in capacity.

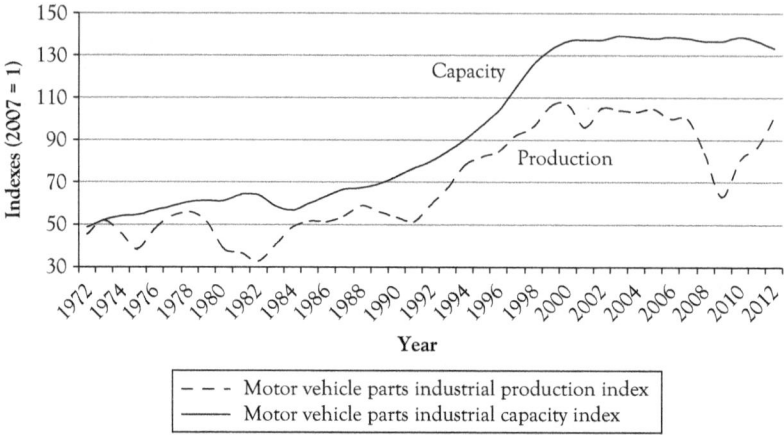

Figure 6.39 Industrial production and capacity indexes for motor vehicle parts

Source: The Federal Reserve (2013).

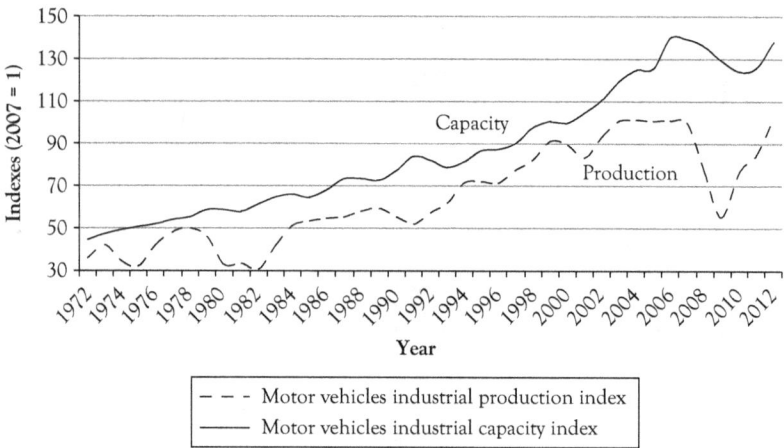

Figure 6.40 Industrial production and capacity indexes for motor vehicles

Source: The Federal Reserve (2013).

Figure 6.41 presents the IP index and the industrial capacity index for electrical equipment, appliances, and components. The pattern seen before is even more dramatic for this industry. As may be seen, production and capacity grow until the year 2000 then both begin a fairly steep decline. This is a U.S. industry in retreat—it has no relationship with the agricultural model. Output is declining and, with it, employment in the United States is declining. Figure 6.42 presents the BOP for this industry. In 1990,

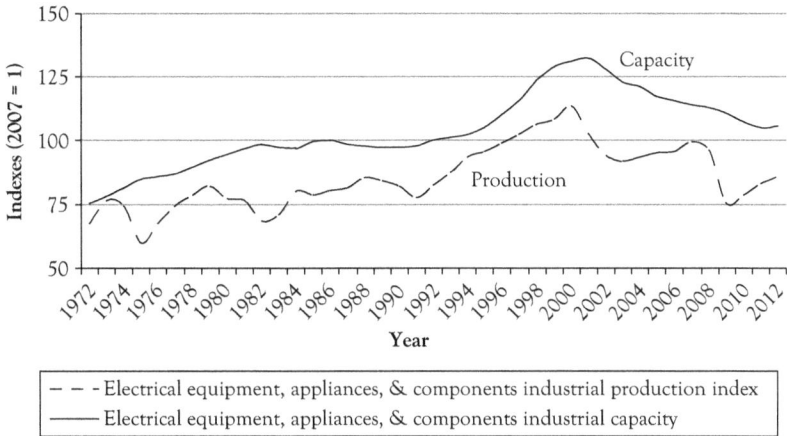

Figure 6.41 Industrial production and capacity indexes for electrical, equipment, appliances, and components

Source: The Federal Reserve (2013).

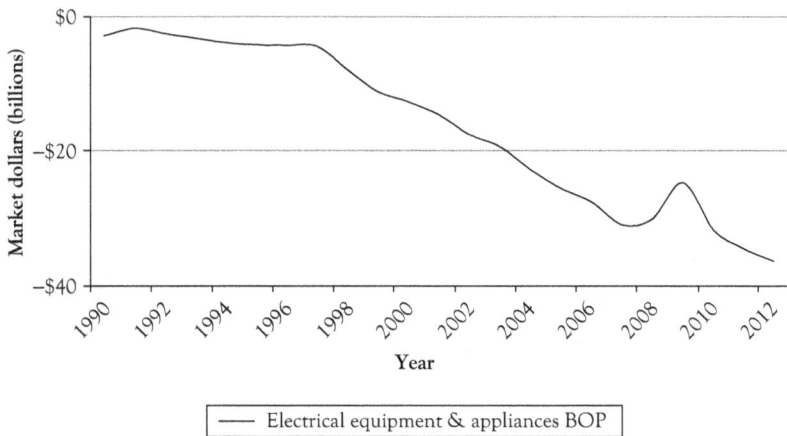

Figure 6.42 Balance of payments for electrical, equipment, appliances, and components

Source: ITA Trade with Selected Market (2014).

the BOP was a negative $3 billion, by 1995 that had grown to $4 billion, by 2000 the BOP had grown rapidly to negative $12.6 billion, by 2005 it had doubled to negative $25.5 billion, and in 2007 it was a negative $31 billion. Little wonder that growth was cut off for the U.S. manufacturers.

As discussed in the section Measuring Output for Manufacturing Industries—It's Harder Than It Looks, measurement of real output for computer and electronic products is confusing, if not tricky. As Figure 6.43 shows, the

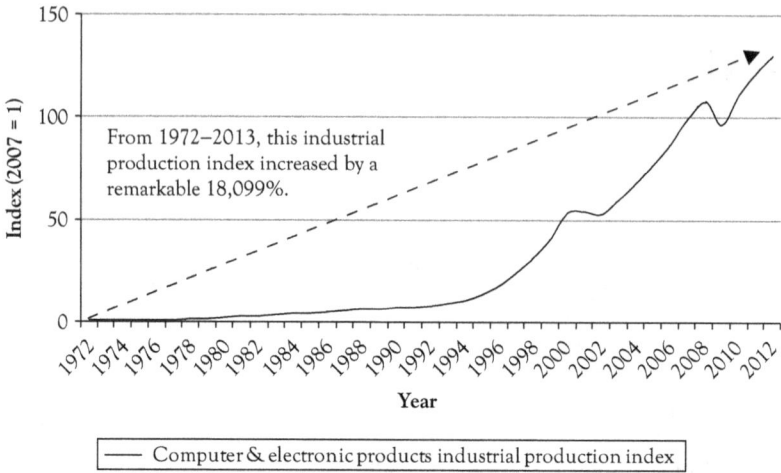

Figure 6.43 Industrial production index for computer and electronic products

Source: The Federal Reserve (2013).

IP index for this sector increased by a staggering 18,099 percent between 1972 and 2012. As noted earlier, most of this growth is due not to an increase in units shipped but the rapidly advancing capabilities of the products.

This sector is composed of five major industries. Table 6.7 presents the total change in the production index for the sector and the five key industries. Two of the industries that create the overall very high growth rates are computer and peripheral equipment and semiconductors and other electronic components. Computer and peripheral equipment increased by 144,848 percent and semiconductors and other electronic components by 224,777 percent. Clearly neither of these industries are currently that much larger than they were in 1972. The other three industries (communications equipment, audio and video equipment, and navigational, measuring instruments, etc.) have much smaller growth rates. In fact the audio and video equipment industry has negative growth. It is important to note that technological capabilities are not factored in the estimation for the changes in the IP indexes for these three industries so the explosive growth is not seen.

Figure 6.44 presents the IP index for computer and peripheral equipment. Even with the dramatic drop due to the recession, this index still increased by 144,848 percent between 1972 and 2013.

As shown in Figure 6.45, however, the value of shipments (VS) for U.S. manufactured products in this industry (in current market dollars)

Table 6.7 Industrial production (IP) index: Overall percentage change and average annual cumulative growth rates (AACGR) for computer and electronic products

	IP Index for Computer & Electronic Products	IP Index for Computer & Peripheral Equipment	IP Index for Communications Equipment	IP Index for Audio & Video Equipment	IP Index for Semiconductor & Other Electronic Components	IP Index for Navigational, Measuring, Electromedical, & Control Instruments
1972–2012	18,098.57%	144,847.74%	2,713.16%	-2.18%	224,777.05%	710.15%
AACGR	13.53%	19.43%	8.48%	-0.05%	20.71%	5.23%

Source: The Federal Reserve (2013).

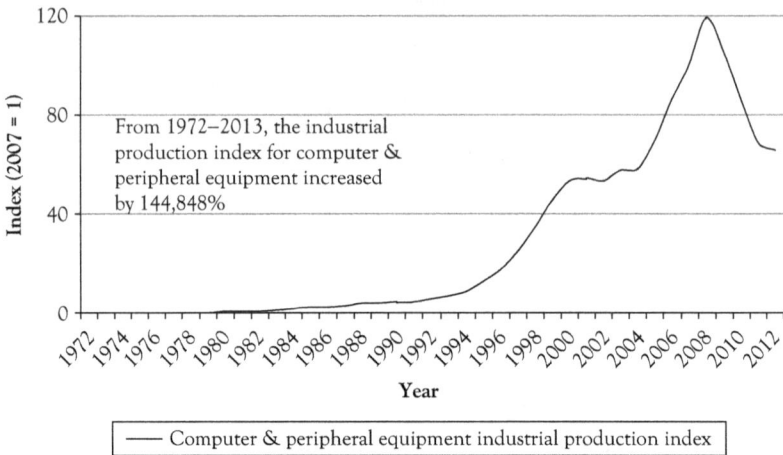

From 1972–2013, the industrial production index for computer & peripheral equipment increased by 144,848%

—— Computer & peripheral equipment industrial production index

Figure 6.44 Industrial production index for computer and peripheral equipment

Source: The Federal Reserve (2013).

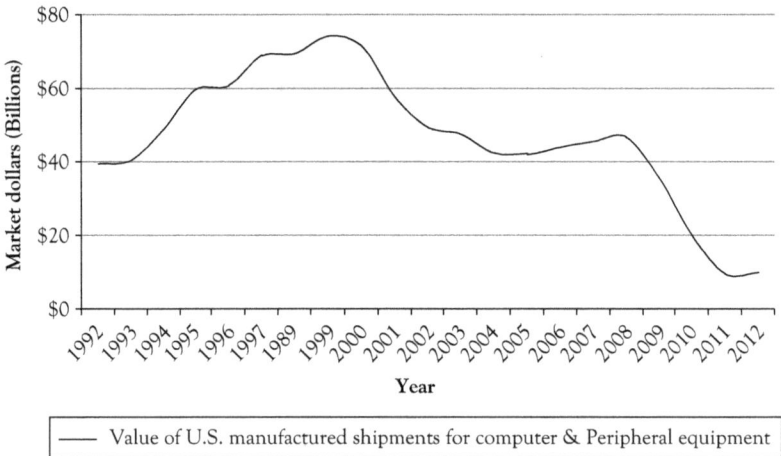

—— Value of U.S. manufactured shipments for computer & Peripheral equipment

Figure 6.45 Value of U.S. manufactured shipments for computer and peripheral equipment

Source: U.S. Census Bureau (2013).

actually declined by roughly 75 percent. Again, this is not the agricultural model of increased output. This is a shrinking industry, albeit one that is shipping faster and more powerful, but fewer, computers.

Figure 6.46 presents the IP index for semiconductors and other electronic components. As may be seen, this index increased by 224,777 percent between 1972 and 2013. Again this reflects the impacts of Moore's Law (Figure 6.7)

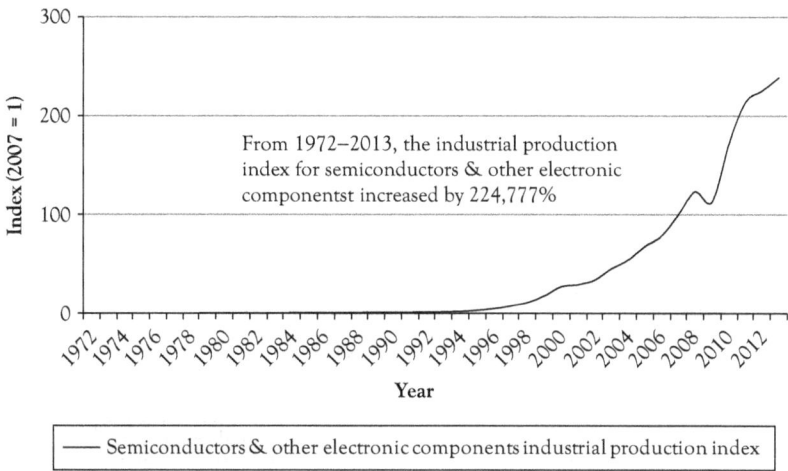

From 1972–2013, the industrial production
index for semiconductors & other electronic
componentst increased by 224,777%

— Semiconductors & other electronic components industrial production index

**Figure 6.46 Industrial production index for semiconductors and other
electronic components**

Source: The Federal Reserve (2013).

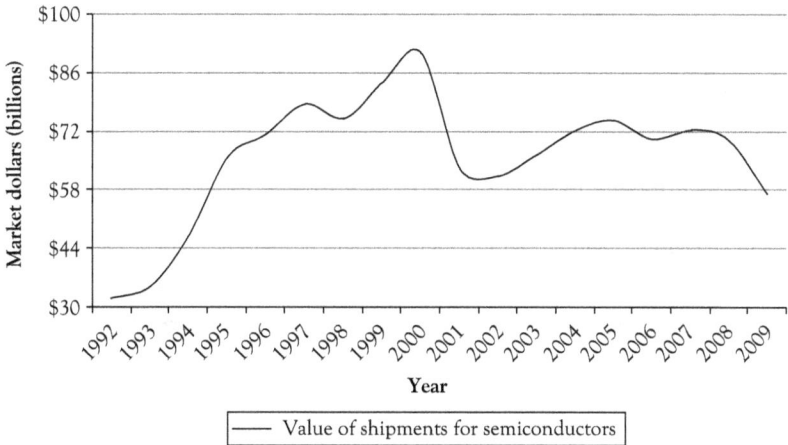

— Value of shipments for semiconductors

**Figure 6.47 Value of shipments for U.S. manufactured
semiconductors**

Source: U.S. Census Bureau (2013).

that shows that the number of transistors on an integrated circuit will double
every two years. Other performance capabilities such as speed and memory
capacity are strongly linked with Moore's Law. Nevertheless, even with such
astounding growth in the production index, gross output (sales) in current
dollars was up only 100 percent over 41 years, as shown in Figure 6.47. This
is not a rapid growth in production or sales but capabilities of the product.

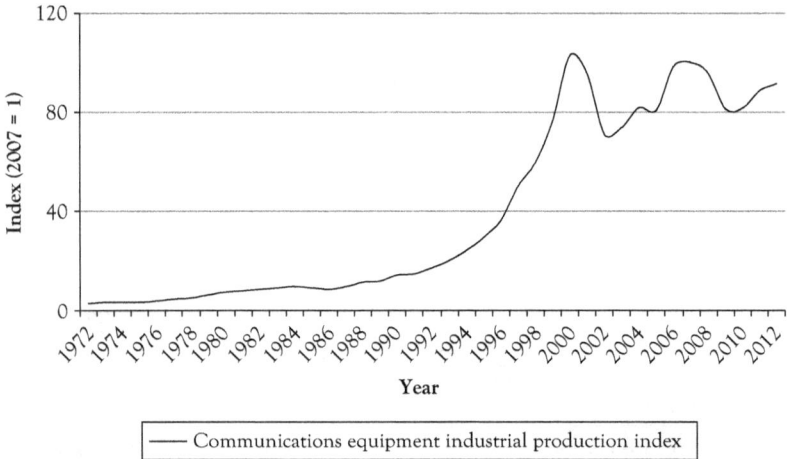

Figure 6.48 Industrial production index for communications equipment

Source: The Federal Reserve (2013).

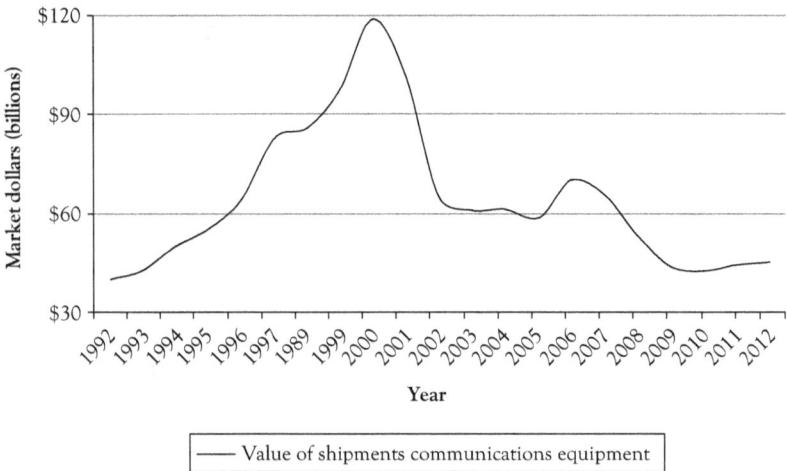

Figure 6.49 Value of shipments for U.S. manufactured communications equipment

Source: U.S. Census Bureau (2013).

Figure 6.48 presents the IP index for manufacture of communications equipment. As may be readily seen, this production grew significantly in the 1980s and 1990s but then peaked in 2000 and has remained below that level, more or less flattening out. Gross output or sales, however, shows a much worse trend in Figure 6.49, with sales peaking in 2000 and then experiencing a decline of roughly 65 percent, falling from $120 billion to $40 billion. Again, this is not agriculture with increasing output. This is a U.S. manufacturing industry in decline, with employment and establishments falling with increased output. The Defense Production Act Committee formed a telecommunications study group to address critical concerns regarding the state of the U.S. telecommunications industry and the impact it has on national security and economic competitiveness. The study group found, not surprisingly, "alarming trends in the health of U.S. telecommunications companies" (Office of the Deputy Assistant Secretary of Defense 2013). The study found that the United States has only one domestic firm in the top tier, a few medium-size manufacturers (annual sales exceeding $500 million), and several smaller vendors. The United States no longer has a wireless equipment vendor capable of producing at scale. Three primary consequences of the market's transformation include:

1. The United States is losing its capabilities in key equipment sectors;
2. There are fewer leading U.S. vendors for agencies and universities to partner with for research and development (R&D); and
3. The options to successfully translate domestic innovation into U.S. telecommunications equipment are increasingly limited.

According to Manufacturing & Technology News, "Those involved in the study say that the government is now presented with a vexing challenge. Since there is hardly anything left to salvage of the U.S. telecommunications hardware industry, the only option available is reinventing the U.S. industry by investing in new technologies, such as photonic integrated circuits, that could form the basis of a new industrial capacity" (McCormack 2014).

Figure 6.50 presents the IP index for the U.S. manufacturing of audio and video equipment. This index grew in the 1990s but then fell sharply in the recession with only a modest recovery. In contrast, sales were flat and then fell by 80 percent. In fact, who knew audio and video equipment was even made in the United States anymore. With this sales trend, it probably won't be.

For this industry, since there is no adjustment for technological capabilities, the value of shipments in Figure 6.51 has a much more direct relationship with the IP index.

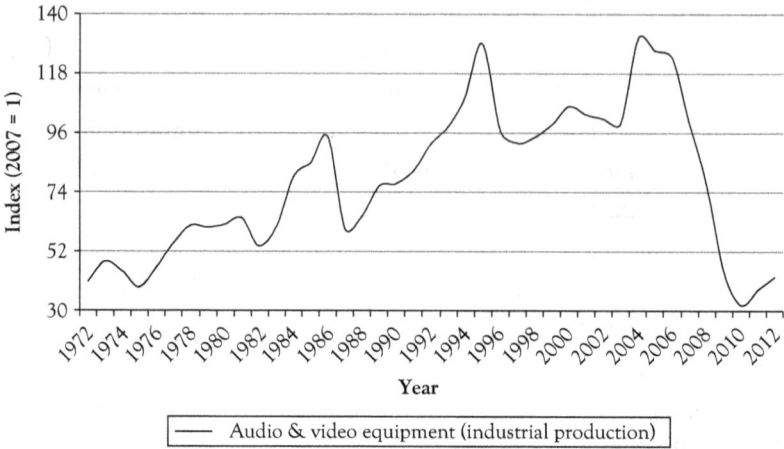

Figure 6.50 Industrial production index for audio and video equipment

Source: The Federal Reserve (2013).

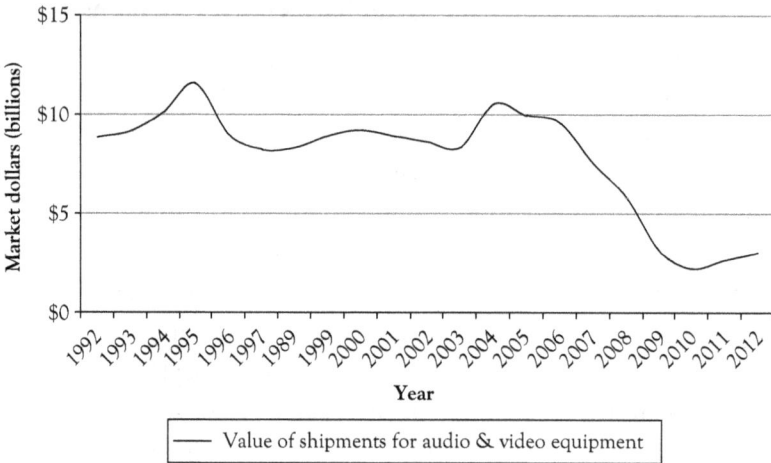

Figure 6.51 Value of shipments for U.S. manufactured audio and video equipment

Source: U.S. Census Bureau (2013).

The final industry in this sector is for navigational, measuring, electro-medical, and control instruments. As seen in Figures 6.52 and 6.53, both the production index and the VS have been growing over this period, however, the production index has grown much more than sales, tending to overstate the health of this industry.

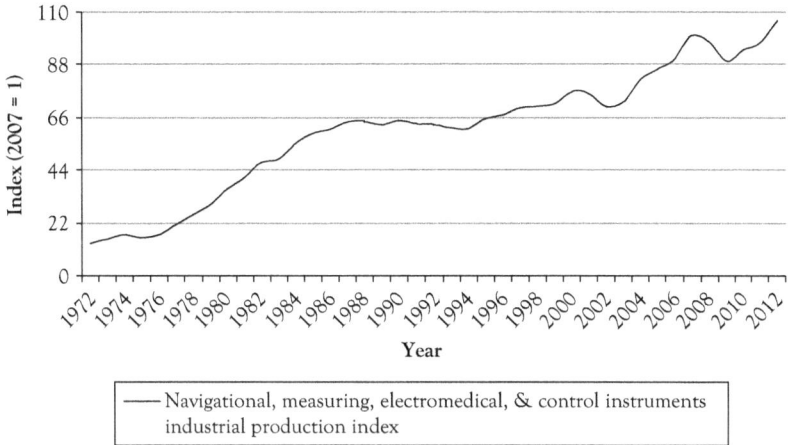

Figure 6.52 Industrial production index for navigational, measuring, electromedical, and control instruments

Source: The Federal Reserve (2013).

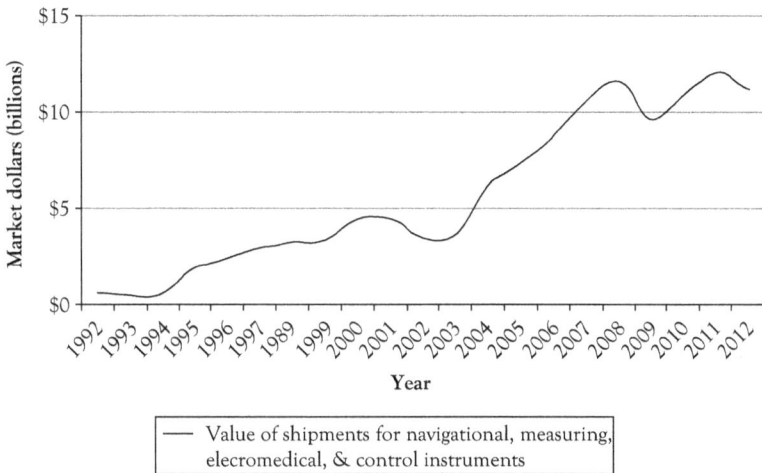

Figure 6.53 Value of shipments for U.S. manufactured navigational, measuring, electromedical, and control instruments

Source: U.S. Census Bureau (2013).

Summary Regarding Industrial Output

It has been shown by comparing the IP indexes for both computers and peripheral products and semiconductors with their respective sales that the estimates for these two industries overstate the growth and health

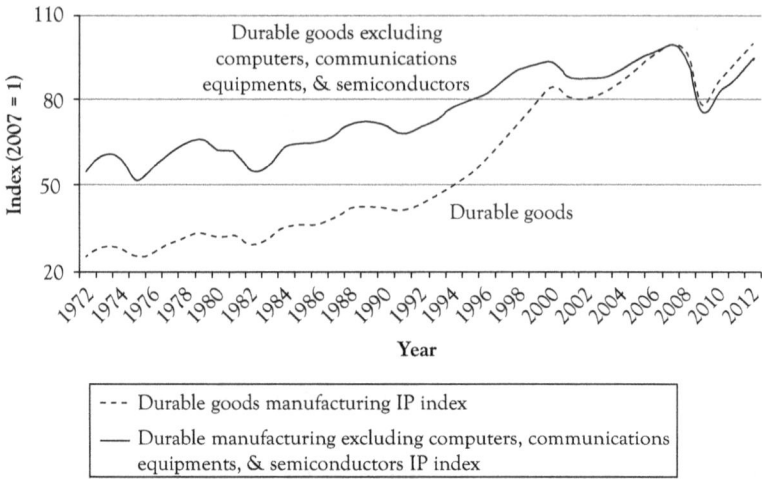

Figure 6.54 Industrial production indexes for durable goods, and durable goods excluding computers, communications equipment, and semiconductors

Source: The Federal Reserve (2013).

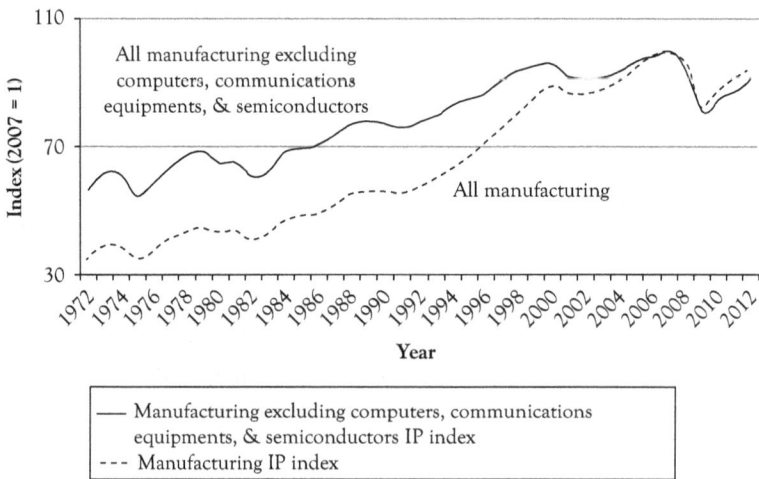

Figure 6.55 Industrial production indexes for all manufacturing and all manufacturing excluding computers, communication equipment, and semiconductors

Source: The Federal Reserve (2013).

of these industries. These overstatements are then reflected in the higher level statistic for durable goods and, thus, in the statistic for all of U.S. manufacturing. Figure 6.54 illustrates the dramatic difference. Including computers, and so forth, the production index for durable goods increased by 292 percent in the 40 years from 1972–2012, that is a healthy 3.5 percent growth per year. However, excluding computers, semiconductors, and so forth, durable goods only increased by 72 percent or 1.4 percent a year, a much weaker industry and growth.

Figure 6.55 shows this impact for all of U.S. manufacturing. The IP Index increased by 164 percent or 2.45 percent a year over these 40 years, a very respectable growth rate. However, removing the estimates for computers, semiconductors, and so forth, manufacturing then only grew by 60 percent or 1.18 percent a year, a much weaker and more negative situation. Chapter 7 addresses a key factor behind the true state of American manufacturing.

Key Takeaways

Manufacturing in the United States is struggling. Production and productivity is not soaring—it's largely an illusion. Many industries have been more or less decimated. Stagnant and declining sales are not uncommon. Large job losses and plant closings are common across industries. The official government statistics are too easily misinterpreted to show growing productivity and output. That growth is largely an illusion created by an attempt to capture technological advancements in a single sector of manufacturing—computers and electronic products. IP, value of shipments, and industrial capacity have stagnated or declined for many manufacturing industries after the year 2000.

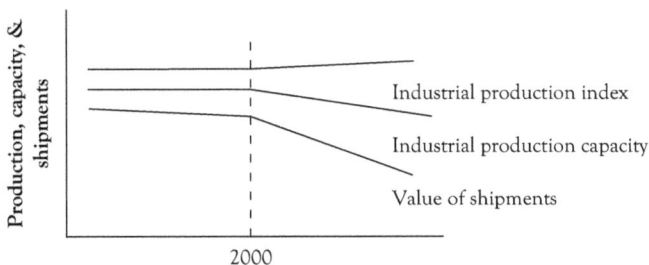

Data pattern 4. Industrial production index, capacity index, and value of shipments

CHAPTER 7

Imports and Balance
of Payments

Introduction

The United States is appropriately concerned about the highly negative BOP for imported oil. That trade deficit is now running a little over $300 billion each year. What is not often appreciated is that the U.S. negative BOP for manufactured goods has generally *vastly exceeded* the deficit for oil. As shown in Table 7.1, in the year 2000, the BOP for manufactured goods was negative $316 billion while that for oil was negative $104 billion—the deficit for manufactured goods exceeded that of oil by $212 billion. With the increasing price of oil, the gap has closed somewhat but in 2012, the deficit for manufactured goods was negative $457 billion compared to negative $315 billion for oil. Importing these manufactured goods into the United States rather than making them here is a serious issue for the economy and national security.

Figure 7.1 presents the BOP over time for manufactured goods and oil. The BOP for manufactured goods peaked in 2006 at a deficit of $567 billion while the BOP for oil peaked in 2008 at a deficit of $384 billion.

Table 7.1 Balance of payments for manufactured goods compared to oil

(Billions of Dollars)						
	1990	**1995**	**2000**	**2005**	**2007**	**2012**
Balance of payments for manufactured goods	($77.971)	($130.083)	($316.065)	($541.373)	($542.086)	($457.075)
Balance of payments for oil & gas	($48.300)	($49.125)	($104.605)	($217.903)	($276.010)	($315.100)

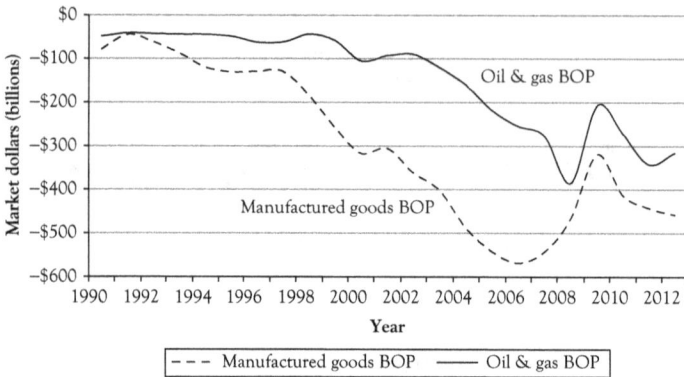

Figure 7.1 **Balance of payments for manufactured goods and for oil and gas**

Source: ITA Trade with Selected Market (2014).

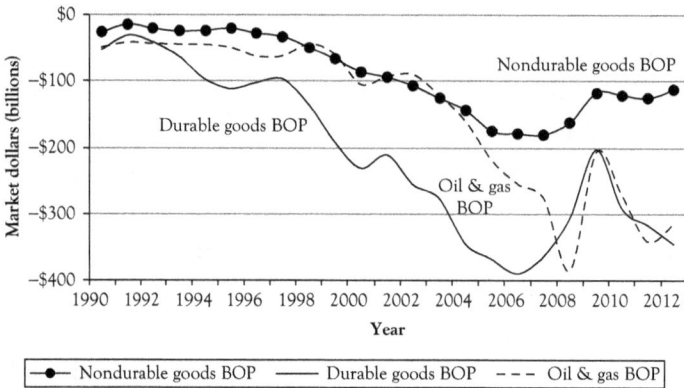

Figure 7.2 **Balance of payments for durable and nondurable goods and for oil and gas**

Source: ITA Trade with Selected Market (2014).

With growing U.S. oil production, the BOP for oil is now being reduced while the U.S. deficit for manufactured goods is $180 billion higher and continues to grow steadily higher and higher.

Figure 7.2 shows that, over time, the BOP for durable goods had generally exceeded that for oil and that, until 2004, the deficits for oil and for nondurable goods (apparel, and so forth) were roughly the same. As may be seen, in 1990, the BOP for durable goods was a negative $52.4 billion and the BOP for oil was negative $48.3 billion. By 1995, the negative BOP for durable goods had doubled in just five years to a deficit of

$111 billion with oil only worsening to a negative $49 billion. By 2000, the durable goods BOP was a negative $230 billion, again more than doubling in a five-year period. Oil BOP also doubled between 1995 and 2000 and reached a negative $104 billion yet that was $126 billion less than that for durable goods. By 2005, the deficit for durable goods exceeded that of oil by a remarkable $150 billion and maintained roughly that level until the Great Recession. Imports of durable goods fell substantially in the recession and fell earlier than the imports of oil. As a result, the deficit for oil briefly exceeded that for durable manufactured goods, but by 2012 the deficit for manufactured goods exceeded that for oil. With increased U.S. production and greater efficiencies, the deficit for oil has actually begun to go down while that for durable manufactured goods continues to climb.

Balance of Payments for Durable Goods Industries

Table 7.2 presents the durable goods manufacturing industries that have the ten largest BOP deficits. The motor vehicles industry has the largest deficit but it has been relatively stable since the year 2000. The second, third, sixth, and ninth industries with the largest deficits are all within the computers and electronic products subsector, specifically, communications equipment, computers and peripheral equipment, audio and video equipment, and semiconductors. As shown in Figure 7.3, over the period from 1990 to 2012, the trade deficit for computers and electronic product soared from a negative $3.65 billion to a negative $151.3 billion. Meanwhile, as may be seen in Figure 7.4, the value of shipments for U.S.-manufactured products in this subsector peaked in the year 2000 and then fell by roughly $150 billion. The similarity of these two figures is striking—shipments of U.S.-made products down $150 billion and trade deficit worsening by $150 billion.

Figure 7.5 presents the BOP trends for industries within the computer and electronic products subsector. As may be seen, three of the industries have seen rapid deterioration in their BOP with more and more products coming into the United States. Only the electronics instruments manufacturing industry has a positive BOP. It would be hard to imagine that these large trade deficits would not have impacts on the shipments of manufactured goods by U.S. manufacturers.

Table 7.2 Top ten durable goods manufacturing industries with largest deficits in balance of payments (billions of market dollars)

Industry	1990	1995	2000	2005	2007	2010	2012
Motor vehicles	–$41.6	–$52.9	–$101.6	–$101.2	–$96.2	–$76.9	–$97.4
Communications equipment	–$2.5	$0.6	–$9.5	–$28.5	–$42.2	–$52.7	–$59.4
Computer & peripheral equipment	$10.9	–$1.4	–$13.7	–$33.0	–$37.7	–$45.0	–$51.3
Motor vehicle parts	–$5.0	–$4.8	–$5.9	–$26.8	–$28.3	–$25.5	–$39.3
Other miscellaneous manufacturing	–$14.3	–$19.8	–$34.9	–$41.0	–$43.7	–$37.7	–$32.3
Audio & video equipment	–$9.4	–$14.5	–$23.1	–$34.0	–$39.9	–$35.4	–$29.4
Iron & steel mills & ferroalloy production	–$6.2	–$8.5	–$11.4	–$17.2	–$20.5	–$12.3	–$21.1
Household & institutional furniture	–$2.3	–$3.8	–$9.9	–$17.1	–$18.3	–$16.1	–$17.5
Household appliances	–$1.7	–$2.1	–$4.9	–$11.7	–$14.4	–$15.3	–$16.8
Semiconductors & electronic components	–$7.1	–$22.1	–$14.9	–$5.6	$1.5	–$5.3	–$13.9
Electrical equipment	$0.0	–$1.1	–$3.8	–$5.6	–$6.4	–$6.4	–$8.1

Source: ITA Trade with Selected Market (2014).

Figure 7.6 presents the trend in the value of shipments (sales) for U.S.-manufactured computers and peripheral products with the BOP for those products. As may be seen in the figure, the value of shipments of U.S.-manufactured computers and peripheral products peaked at $74 billion in the year 1999 and has fallen by approximately $60 billion to only $10 billion in 2012. During that same period the BOP for these products soared to approximately a negative $50 billion deficit. As Atkinson warns, "It appears that U.S. manufacturing now experiences what can be called a one-way job loss ratchet, with significant job losses in economic downturns but

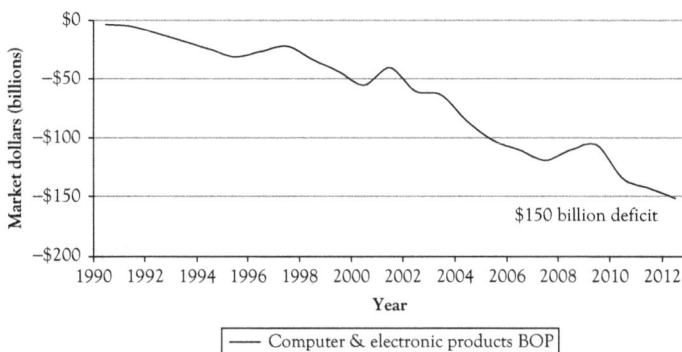

Figure 7.3 Balance of payments for computer and electronic products

Source: ITA Trade with Selected Market (2014).

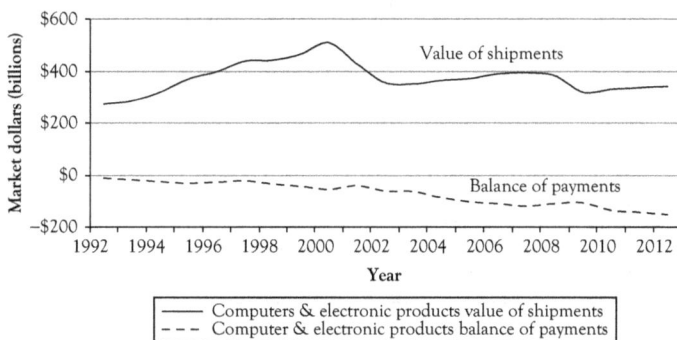

Figure 7.4 Balance of payments for computers and electronic products compared to the value of shipments of U.S.-manufactured products

Sources: ITA Trade with Selected Market (2014); U.S. Census Bureau (2013).

then very shallow job gains, if any, in the recovery period." The U.S. computer and peripheral equipment manufacturing industry is at great risk and this poses huge risks for U.S. national security as industry, government, and households become ever more digitally driven. The dangers of embedded malware and counterfeit products can only grow higher and higher. A 2011 Senate Armed Services Committee investigation found at least 1,800 cases of counterfeit parts in U.S. weapons and about 1 million suspected counterfeit parts in the supply chain. In a single missile interceptor system, the Missile Defense Agency found 800 fake parts; it cost over $2 million to replace them (U.S. Senate Committee on Armed Services 2011).

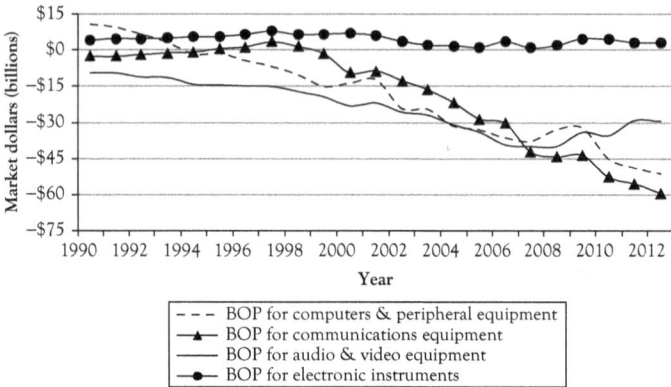

Figure 7.5 *Balance of payments for industries in the computers and electronic products manufacturing sector*

Source: ITA Trade with Selected Market (2014).

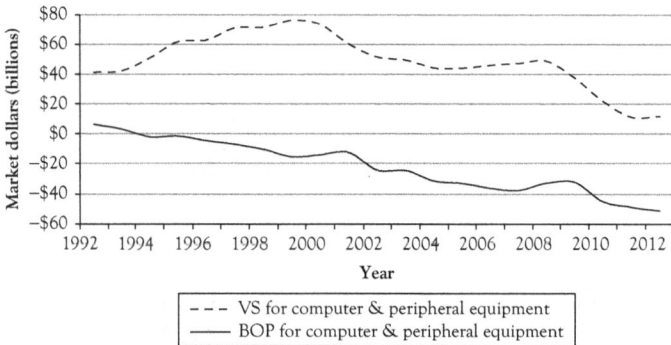

Figure 7.6 *Comparison of the value of shipments of U.S.-manufactured computer and peripheral equipment with corresponding balance of payments*

Source: U.S. Census Bureau (2013); ITA Trade with Selected Market (2014).

Figure 7.7 presents U.S. imports of computers and peripheral equipment by exporting country. From 1990 to 2000, Japan was the leading exporter of these products to the United States, and Taiwan was the second largest exporter to the United States. In 2000, China was only exporting about $8 billion of computers and peripheral equipment to the United States. By 2012, imports from China had soared to roughly $65 billion and, meanwhile, imports from Japan, Taiwan, and Malaysia have nearly gone to zero. Only Mexico has been able to maintain a significant level of exports of computers and peripherals to the United States. China

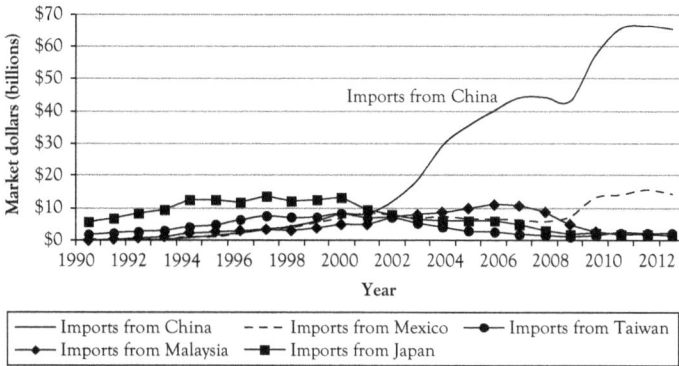

Figure 7.7 U.S. imports of computers and peripheral products by leading countries

Source: ITA Trade with Selected Market (2014).

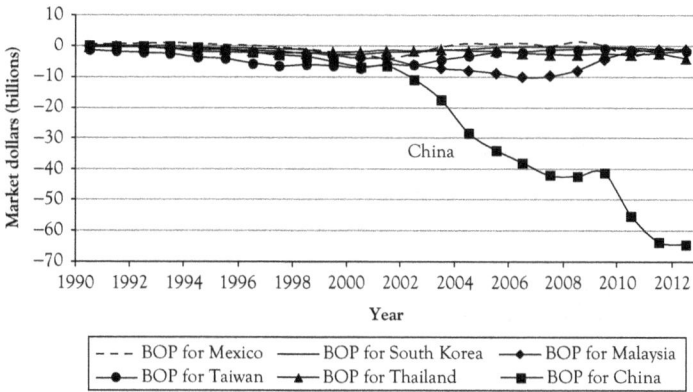

Figure 7.8 Balance of payments for computers and peripheral products for countries where the United States has the largest deficit

Source: ITA Trade with Selected Market (2014).

appears to be on its way to global dominance in this critical industry, if not a global monopoly.

Figure 7.8 presents the BOP for computer and peripheral products for the six countries with which the United States has the largest deficits. As may be seen, reflecting the huge growth in imports from China, the BOP with China has gone from a deficit of $5 billion in 2000 to a deficit of roughly $65 billion in 2012 for computers and peripheral equipment alone. The U.S. deficits for the other five countries are all under $5 billion.

Table 7.3 presents the historical data for the BOP for computers and peripheral equipment. In 1990, the largest deficit was with Japan at $2.35 billion. By 2000, the U.S. deficit with Japan in this industry had grown to $8.22 billion. After 2000, the deficits for computers from Japan began to rapidly decline. A similar pattern may be seen for Malaysia. After the year 2000, the BOP with China exploded, displacing shipments of computers and peripheral equipment made in the United States and in other countries.

Figure 7.9 presents the trend for the value of shipments (sales) of U.S.-manufactured communication equipment and the trend for the BOP for

Table 7.3 U.S. trade deficits for computers and peripheral equipment (billions of current dollars)

	1990	1995	2000	2005	2010	2013
Japan	–$2.35	–$7.96	–$8.22	–$3.79	–$0.97	–$0.42
South Korea	–$0.41	–$0.70	–$2.85	–$0.64	–$0.73	–$0.88
Malaysia	$0.08	–$2.03	–$4.07	–$8.96	–$1.98	–$1.21
Viet Nam	$0.00	$0.02	$0.01	–$0.07	–$0.20	–$1.61
Taiwan	–$1.37	–$4.22	–$7.15	–$2.23	–$1.19	–$1.79
Thailand	–$0.32	–$1.29	–$1.98	–$1.86	–$2.88	–$4.45
China	$0.12	–$1.23	–$6.78	–$33.70	–$55.00	–$63.42

Source: ITA Trade with Selected Market (2014).

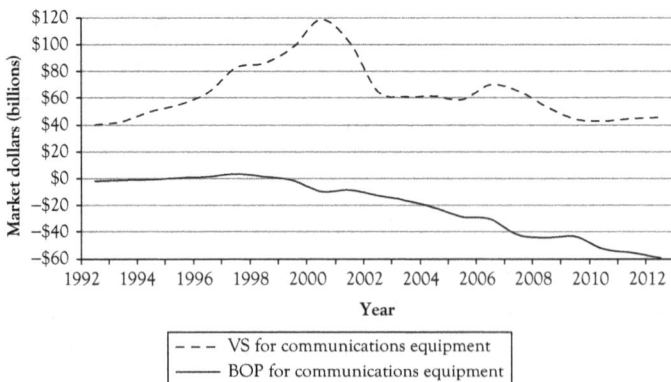

Figure 7.9 Comparison of the value of shipments of U.S.-manufactured communications equipment with corresponding balance of payments

Source: U.S. Census Bureau (2013); ITA Trade with Selected Market (2014).

those products. Once again the same pattern is seen as in Figure 7.6. The value of U.S.-manufactured products shipped peaked in 2000 at $119 billion but then plummeted to $46 billion by 2012. Over the same period, the trade deficit for these products soared from a negative $10 billion to a negative $60 billion. As was seen for computers and peripheral equipment, U.S.-manufactured communication equipment is being displaced by imported products. Figure 7.10 presents the trends over time for U.S. imports of communication equipment. Again, the pattern seen in Figure 7.7 for computers is seen again here for communication equipment. Imports from China exploded after the year 2000 and shipments from other countries struggled to maintain any market share. Figure 7.11 presents the BOP for communication equipment for the six countries with which the United States has the largest trade deficits for this equipment. Again the situation is very similar to that seen in Figure 7.8. The U.S. deficit with China in this industry has soared from roughly negative $3 billion in 2000 to a $50 billion deficit in 2012. Four of other five leading countries are below $5 billion, with Mexico around $8 billion. Table 7.4 presents historical data for the U.S. deficits related to the imports of communication equipment. As with computers, in 1990 the largest deficit by far for communications equipment was with Japan. That was rapidly diminished as the deficit with China exploded. This is another

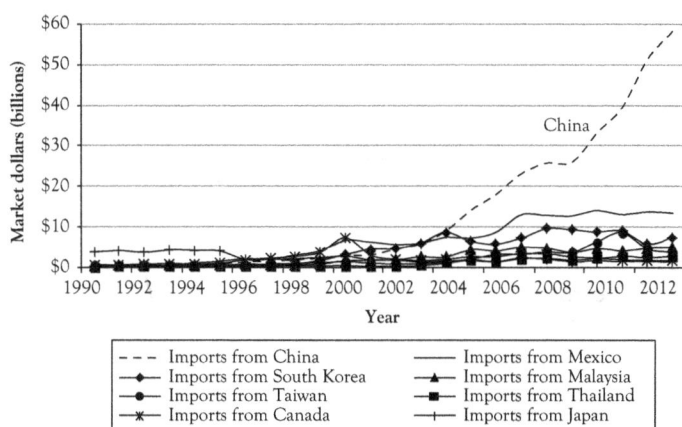

Figure 7.10 U.S. imports of communication equipment by leading countries

Source: ITA Trade with Selected Market (2014).

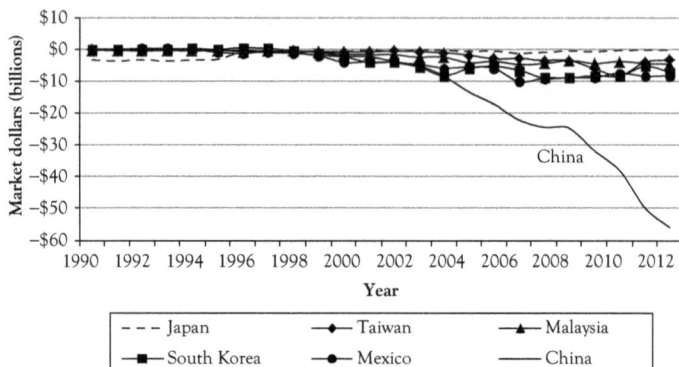

Figure 7.11 *Balance of payments for communication equipment for the countries where the United States has the largest deficit*
Source: ITA Trade with Selected Market (2014).

Table 7.4 *U.S. deficit for communication equipment (billions of current dollars)*

	1990	1995	2000	2005	2010	2013
Japan	–$3.39	–$3.12	–$0.72	–$0.63	–$0.70	–$0.26
Indonesia	$0.03	$0.13	–$0.05	–$0.20	–$0.33	–$0.30
Poland	$0.00	$0.01	$0.06	$0.10	–$0.03	–$0.32
Viet Nam	$0.00	$0.01	$0.01	$0.01	–$0.21	–$0.70
Thailand	–$0.08	–$0.06	–$0.35	–$1.53	–$2.07	–$2.37
Taiwan	–$0.12	–$0.31	–$0.86	–$1.88	–$6.02	–$3.32
Malaysia	–$0.35	–$0.38	–$1.51	–$4.23	–$4.44	–$4.63
South Korea	–$0.15	$0.05	–$2.30	–$5.98	–$8.26	–$7.05
Mexico	$0.06	–$0.94	–$4.13	–$5.55	–$8.77	–$8.23
China	–$0.30	–$0.30	–$2.33	–$13.29	–$31.67	–$55.91

Source: ITA Trade with Selected Market (2014).

critically important industry where China has great global dominance. As the world becomes completely digital, communication equipment will be connecting everything and everyone, and once again raising substantial concerns regarding national security. As with computers, shipments from China are displacing shipments produced in the United States and in other countries.

Figure 7.12 presents the value of shipments for U.S.-manufactured audio and video equipment and the BOP for those products. For this

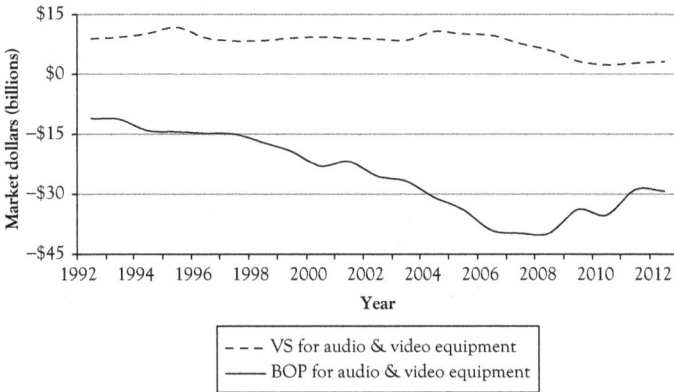

Figure 7.12 Comparison of the value of shipments of U.S.-manufactured audio and video equipment with corresponding balance of payments

Source: U.S. Census Bureau (2013); ITA Trade with Selected Market (2014).

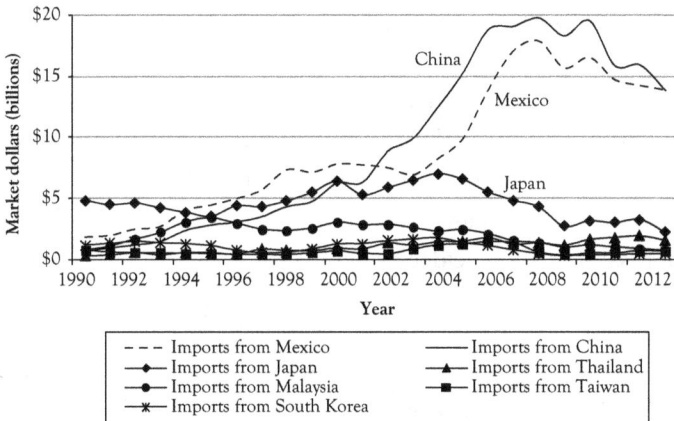

Figure 7.13 U.S. imports of audio and video equipment by leading countries

Source: ITA Trade with Selected Market (2014).

industry in 1990, U.S.-manufactured shipments were about $10 billion and the BOP was about a negative $10 billion. By 2012, U.S.-manufactured shipments had fallen to $3 billion and the BOP had rapidly risen to a deficit of $35 billion. Once again, U.S.-manufactured products have been displaced by imports. For this industry, as shown in Figure 7.13, the largest volume of imports is from China and the second largest is with Mexico. The largest U.S. deficit in 1990 for this industry was once again with Japan. Imports from Japan peaked in 2004 as did the deficit with Japan but then imports from Japan declined rapidly and are now a distant third. Imports from Mexico surged in 2005 and are now about equal to imports from China. Figure 7.14 presents the balance of trade for audio and video equipment for those countries with which the United States has the largest deficits. This figure essentially mirrors the imports shown in Figure 7.13 with the United States having the largest deficits with China and Mexico, and the deficit with Japan peaking in 2004 and then declining. It is interesting to note that both imports and deficits grew rapidly in the past decade but fell during the recession and continued to fall after the recession. A New York Times article "A Bonanza in TV Sales Fades Away" gives three reasons for the strong growth of TV sales in the last decade: (i) the rise of flat-panel television technologies like plasma and LCD; (ii) rapid price reductions (from 2007 to 2010, the average price of an LCD TV dropped 36.3 percent and plasma TV prices dropped 51.6 percent); and (iii) the government-mandated switch over to digital broadcasting and the availability of high-definition shows and movies—something the flat-panel televisions were ready to display (Grobart 2011). Now most homes have flat-panel TVs, and consumers have not adopted 3D TVs and sales of TVs are declining.

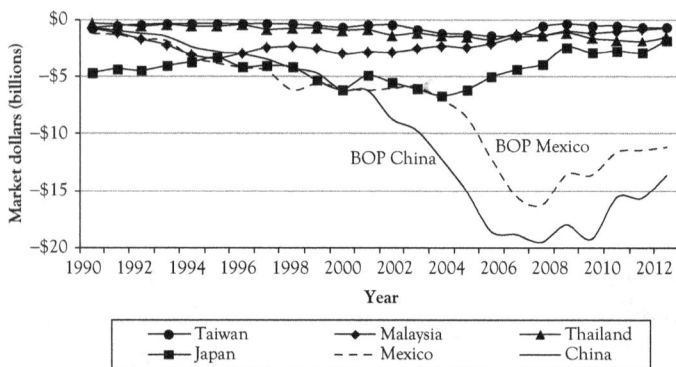

Figure 7.14 Balance of payments for audio and video equipment for the countries where the United States has the largest deficit

Source: ITA Trade with Selected Market (2014).

Figure 7.15 presents the value of shipments by U.S. manufacturers of electronic instruments and the BOP for these products. As may be seen, this is a very different situation. The value of U.S.-manufactured shipments has grown from roughly $80 billion to approximately $130 billion. This is the only industry in this sector with positive growth in sales of U.S.-manufactured products. The BOP for this industry has been small, but positive, ranging from $3 to $4 billion a year. This is an industry where U.S. manufacturers still are globally competitive.

As seen in Table 7.1, the largest trade deficit for manufactured goods is with motor vehicles and parts. Figure 7.16 presents the comparison of the value of shipments of U.S.-manufactured motor vehicles and parts with the corresponding BOP, both in current dollars. As may be seen, the value of shipments of U.S.-manufactured products essentially peaked in 1999 and again reached that level in 2007. Since 1990, however, the trade deficit for motor vehicles and parts has grown steadily reaching a negative $130 billion.

Trade deficits are largely driven by the level of imports. Figure 7.17 presents the cost of imports for both motor vehicles and motor vehicle parts. Currently, the cost of imported vehicles is roughly $170 billion a year and the costs of imported parts are roughly $100 billion. Trade in these two industries will now be examined in more detail.

Figure 7.18 presents the costs of U.S. imports for motor vehicles by the top five exporting countries. A summary of this data is presented in

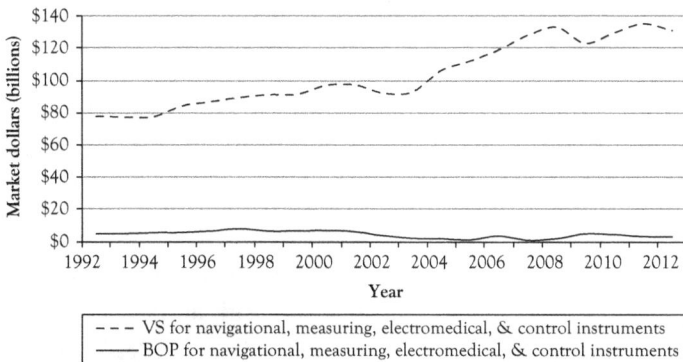

Figure 7.15 Comparison of the value of shipments of U.S.-manufactured electronic instruments with corresponding balance of payments

Source: U.S. Census Bureau (2013); ITA Trade with Selected Market (2014).

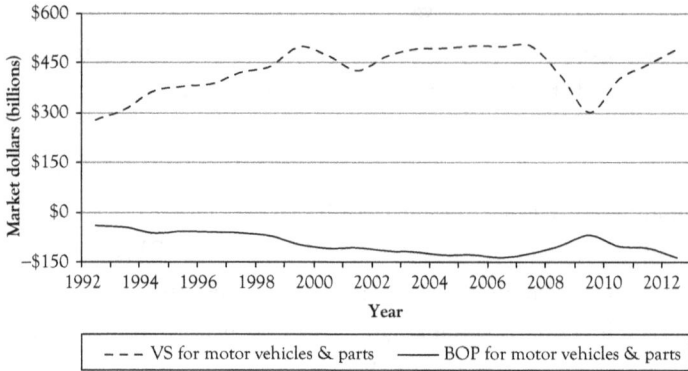

Figure 7.16 Comparison of the value of shipments of U.S.-manufactured motor vehicles and parts with corresponding balance of payments

Source: U.S. Census Bureau (2013); ITA Trade with Selected Market (2014).

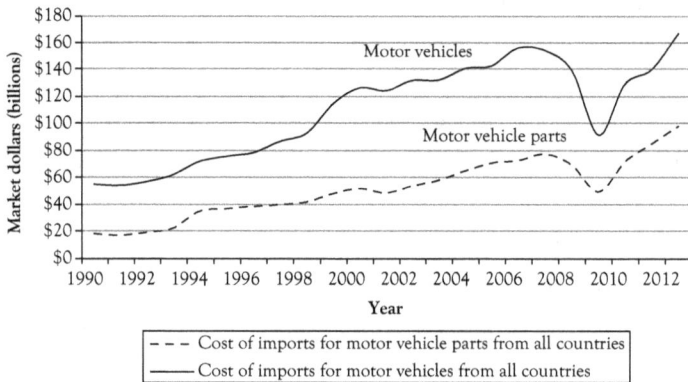

Figure 7.17 Total U.S. imports of motor vehicle and parts

Source: ITA Trade with Selected Market (2014).

Table 7.5. There are several striking trends in this figure. The first is the dramatic growth of imports from Mexico that now exceeds even Japan. Well recognized "American" vehicles such as the Chevrolet Silverado and the Cadillac SUV are now manufactured in Mexico. Even Ram pickups and the Chrysler "Hemi" engines are made in Mexico. Mexico is currently making three million vehicles a year with 80 percent exported to the United States. The output of motor vehicles in Mexico is expected to increase by 38 percent by 2016 (Miroff 2013). GM announced in 2013 that it would invest $691 million to increase its production capacity in Mexico (Reuters 2013). A second striking trend is the stagnation of

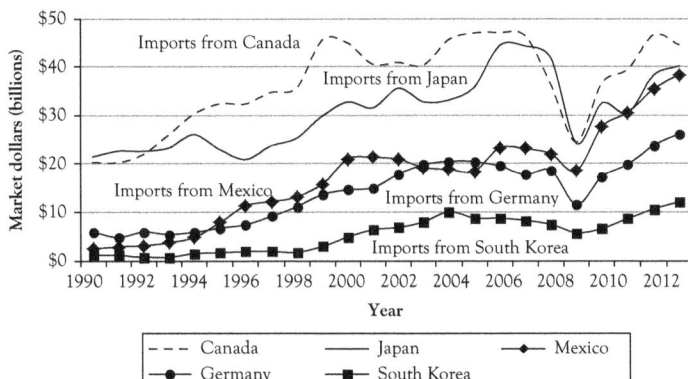

Figure 7.18 *U.S. imports of motor vehicles by leading countries*
Source: ITA Trade with Selected Market (2014).

Table 7.5 Costs of motor vehicle imports by country (billions of current dollars)

	1990	1995	2000	2005	2010	2013
Canada	$20.25	$32.35	$44.86	$46.96	$36.55	$44.51
Japan	$21.45	$22.96	$32.66	$35.76	$32.41	$38.33
Mexico	$2.45	$7.83	$20.99	$18.36	$27.49	$40.10
Germany	$5.88	$6.65	$14.57	$20.38	$17.28	$26.15
South Korea	$1.12	$1.66	$4.87	$8.77	$6.55	$12.04

motor vehicle imports from Canada. Many expect that plants in Canada will be the loser in the competition with Mexico where the hourly wage in the auto plants is $3.20. Between 2000 and 2013, imports from Canada were flat, while during the same period, imports from Mexico doubled from $20 billion to $40 billion. Imports from Germany have grown strongly reflecting increasing demand by American consumers for BMW, Mercedes, and Porsche luxury and performance automobiles.

Figure 7.19 presents the trends for U.S. BOP for motor vehicles for those countries with which the United States has the largest deficits for this industry. The largest U.S. deficits are with Japan and Mexico. Canada has fallen to a distant third while the deficits with Germany and South Korea continue to grow.

As seen in Figure 7.17, the importation of motor vehicle parts has grown rapidly. Figure 7.20 shows that the U.S. trade deficit for vehicle

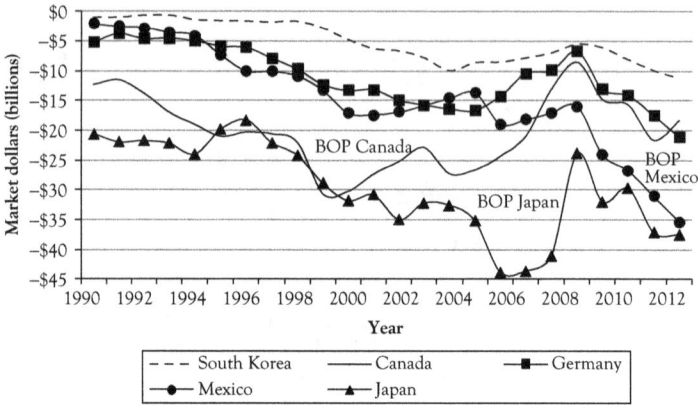

Figure 7.19 Balance of payments for motor vehicles
Source: ITA Trade with Selected Market (2014).

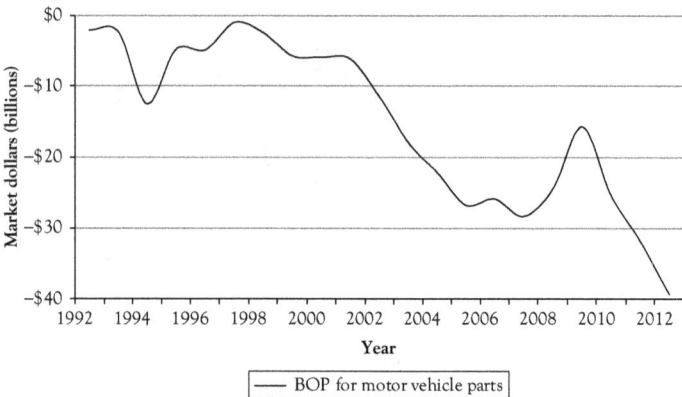

Figure 7.20 Balance of payments for motor vehicle parts
Source: ITA Trade with Selected Market (2014).

parts grew from a negative $2 billion in 1992 to a negative $39 billion in 2012. Unfortunately, the Bureau of the Census does not publish value of shipments for this U.S. industry but it is clearly in danger of being overwhelmed by imports. In the paper "Attack on the American Auto Parts Industry" (Alliance for American Manufacturing 2012), the point is made that "If parts production in the U.S. is falling off, yet overall auto production is increasing, the parts must be coming from somewhere else." Figure 7.21 presents the costs of imports for motor vehicle parts by the leading source countries. Remarkably, imports from Mexico have soared

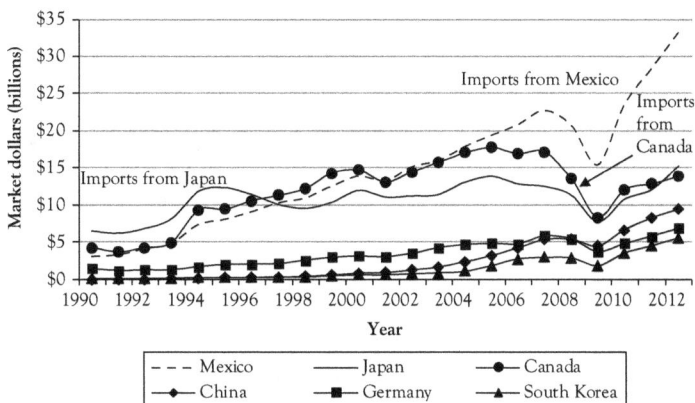

Figure 7.21 U.S. imports of motor Vehicle parts by leading countries
Source: ITA Trade with Selected Market (2014).

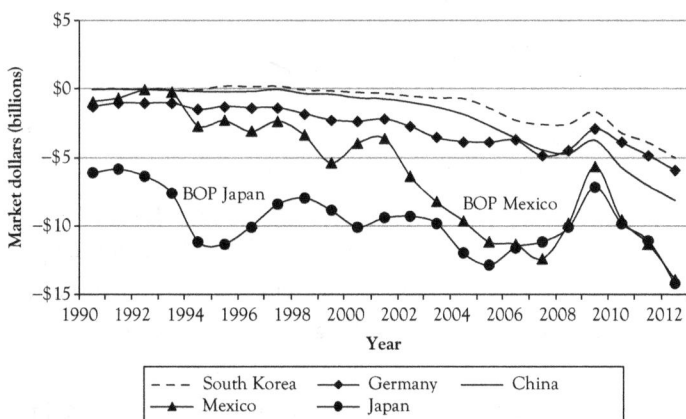

Figure 7.22 Balance of payments for motor vehicle parts for the countries where the United States has the largest deficit
Source: ITA Trade with Selected Market (2014).

from roughly $3 billion in 1990 to nearly $35 billion in 2013. Delphi, formerly part of GM, is one of the largest private employers in Mexico with 54,000 employees working in 46 plants in 22 Mexican cities (Pro Mexico 2014). Figure 7.22 presents the BOP over time for motor vehicle parts for those countries for which the United States has the greatest deficit. The striking aspect of this figure is the growth in the deficit with Mexico from around $3 billion in 1995 to $14 billion in 2012. Mexico and Japan are now about equal with the largest deficits for motor vehicle

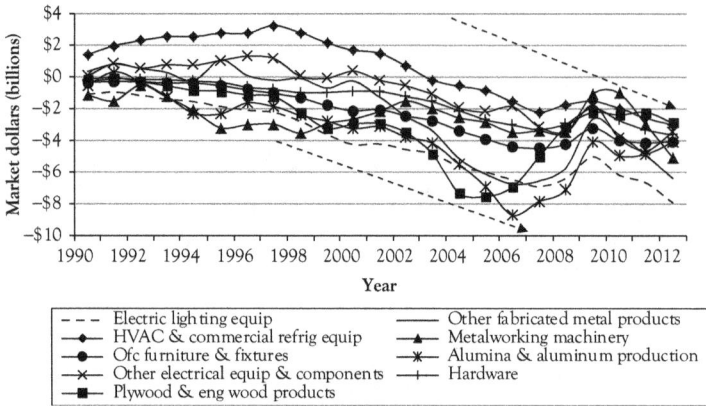

Figure 7.23 Balance of payments for selected durable goods
Source: ITA Trade with Selected Market (2014).

parts. China is third having grown from a $1 billion deficit in 2000 to an $8 billion deficit in 2012. The deficits with all five countries have grown rapidly following the recent recession. Again this most likely indicates continuing loss of market share for U.S. manufacturers.

Figure 7.23 presents the trends for BOP for a number of durable goods. Although it is difficult to identify each industry in the figure, the main intent of this figure is to show the remarkable similarity in the pattern and trends. All have been growing negatively and most have shown substantial growth following the recession. These are products whose importation is slowing the recovery of manufacturing because imported products are taking advantage of the recovery and grabbing increased market share. Growing demand is being met by imports rather than U.S. factories hiring workers.

Balance of Payments for Nondurable Goods

Figure 7.24 presents the trend lines for the U.S. BOP for those nondurable goods with the largest deficits. From 1990 to 2012, the U.S. BOP for apparel grew rapidly from roughly negative $20 billion to negative $70 billion. Once again, as seen in Figure 7.25, this kind of growth in imports cannot occur without substantial impacts on the sales of U.S.-manufactured products. Figure 7.25 presents the sales of U.S.-manufactured

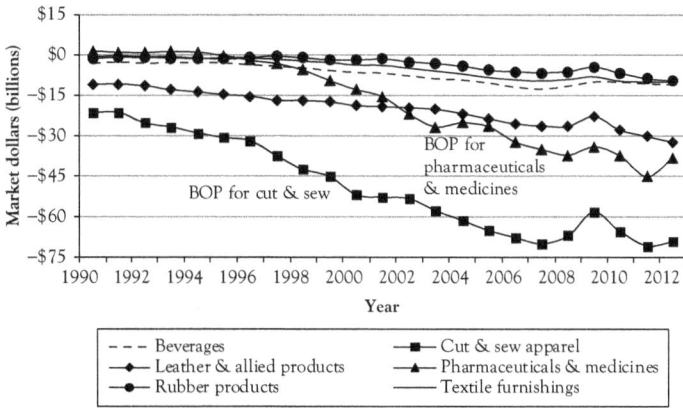

Figure 7.24 Balance of payments for selected nondurable goods
Source: ITA Trade with Selected Market (2014).

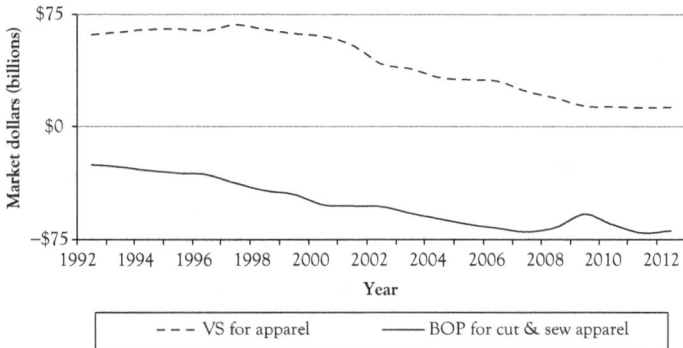

Figure 7.25 Comparison of the value of shipments of U.S.-manufactured apparel with the corresponding balance of payments
Source: U.S. Census Bureau (2013); ITA Trade with Selected Market (2014).

apparel. In 1990 apparel sales were $60 billion but fell rapidly as imports grew, U.S. establishments were closed, and U.S. production dropped. By 2012, sales of U.S.-manufactured apparel had fallen to roughly $13 billion.

A New York Times article sums things up nicely: "The North American Free Trade Agreement (NAFTA) in 1994 was the first blow, erasing import duties on much of the apparel produced in Mexico. The Asian financial crisis in the late 1990's, when currencies collapsed, added a 30 to 40 percent discount to already cheaper overseas products, textile executives said. China joined the WTO in 2001 and quickly became an apparel powerhouse and as of 2005, the WTO eliminated textile quotas. In 1991, American-made apparel accounted for 56.2 percent of all the clothing bought domestically.... By 2012, it accounted for 2.5 percent." Imports have largely displaced apparel made in the United States.

Figure 7.26 presents the BOP for those countries with which the United States has the largest trade deficits for apparel. Again, a very similar pattern over time is seen. On one hand, there are seven countries where the U.S. has deficits of $1 to $7 billion and these deficits have been slowly growing. On the other hand, the trade deficit with China for apparel has exploded going from roughly $4 billion in 1990 to a $32 billion deficit in 2012. As may be seen, the BOP deficit with Mexico began to grow strongly in 1994 with NAFTA but then began to decline in 2001 as China apparel began to take over the market.

As seen in Figure 7.24, the second largest deficit for nondurable goods is for, perhaps somewhat surprisingly, pharmaceuticals and medicines. The BOP for these products has gone from near zero in 1995 to a negative $40 billion in 2012 as seen in Figure 7.24. Over this same period, sales of U.S.-manufactured pharmaceuticals and medicines have grown from $67 billion to $195 billion as shown in Figure 7.27, but the trend line for U.S.-manufactured shipments seems to be leveling as the value of U.S.-manufactured shipments turned down following slight uptick after the recession. These rapidly growing imports have somewhat ominous

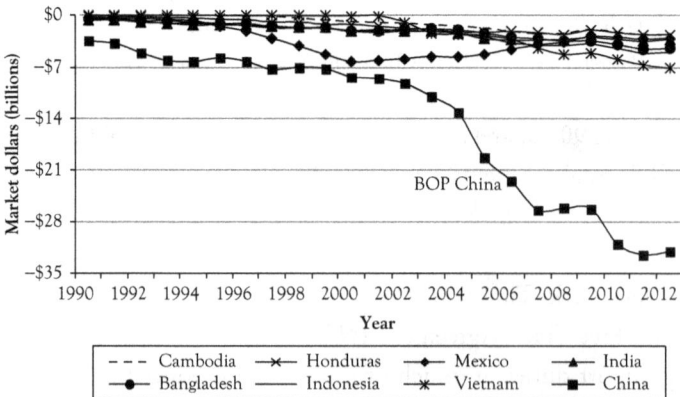

Figure 7.26 *Balance of payments for apparel for the countries where the United States has the largest deficit*

Source: ITA Trade with Selected Market (2014).

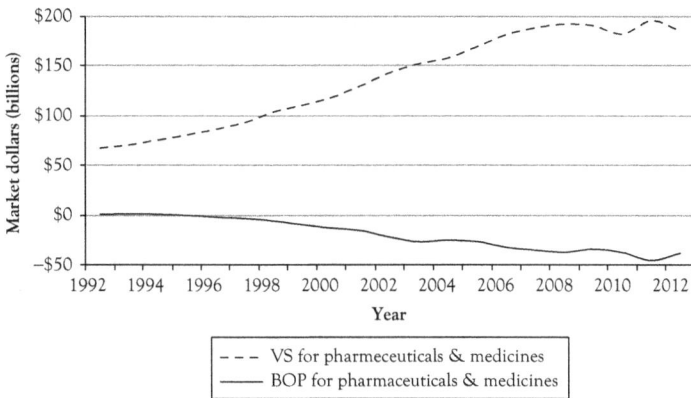

Figure 7.27 Comparison of the value of shipments of U.S.-manufactured pharmaceuticals and medicines with corresponding balance of payments

Source: U.S. Census Bureau (2013); ITA Trade with Selected Market (2014).

implications for U.S. healthcare. In the January of 2008, the FDA began an investigation after receiving reports of serious adverse events in people receiving heparin sodium, a commonly used blood thinner. The agency later learned that an active pharmaceutical ingredient (API) found in this drug contained a contaminant and had been manufactured at a Chinese establishment never inspected by FDA (General Accountability Office 2008, 2011). The FDA attributed 81 deaths to the contaminated heparin and identified nearly 800 individuals with serious injury (CNN 2008). In the 2011 report "Pathways to Global Product Safety and Quality," the FDA estimated that 80 percent of APIs in medications sold in the U.S. are manufactured elsewhere (U.S. Food and Drug Administration 2011). The FDA also noted that half of all medical devices used in this country are imported. In that same report, FDA officials acknowledged that the established system for monitoring and ensuring the quality and safety of drugs could not cope with growing global trade and U.S. imports. India's pharma industry, which supplies 40 percent of the over-the-counter and generic prescription drugs consumed in the United States, is now coming under increased scrutiny by the FDA because of safety lapses, falsified drug tests, and selling fake medicines (Harris 2014).

Another concern related to imports is drug shortages—the United States is experiencing a rapidly increasing frequency of drug shortages as shown in Figure 7.28. Many factors can cause drug shortages (Ventola 2011) but disruptions in the supply of raw or bulk materials are frequently responsible for drug shortages (Ventola 2011). Given that 80 percent of raw materials in the pharmaceuticals sold in the United States are imported, problems at foreign producers can lead to disruptions in the American drug supply. Moreover, the crucial ingredients for nearly all antibiotics, steroids, and other critical drugs are now made exclusively in China (Harris 2014). America's dependency on foreign drug manufacturers appears to be on an ever-worsening path.

Figure 7.29 presents a comparison of the value of shipments of U.S.-manufactured leather goods with the BOP for leather goods. As may be seen, the United States was only manufacturing about $10 billion of these products in 1990 and the balance of trade at that time was approximately a negative $12 billion. By 2012, the value of shipments of U.S.-manufactured leather goods had fallen to $5 billion but the BOP had substantially increased to about $32 billion. Footwear is by far the largest component of leather goods. Figure 7.30 presents the BOP for footwear for the six countries with which the United States has the largest deficits for footwear. Again, the pattern is very similar to other figures of this type. Five countries are in the noise level where the United States has deficits of $1 to $2 billion but, on the other hand, the U.S. deficit with China for footwear has grown rapidly from negative $1 billion in 1990 to negative $17 billion in 2012.

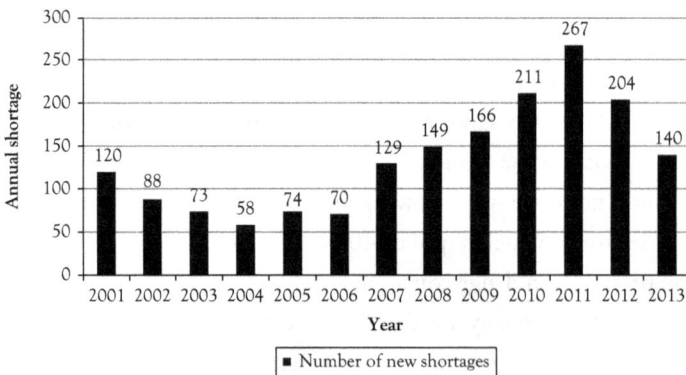

Figure 7.28 National drug shortages 2001–2013

Source: American Society of Health-System Pharmacists (2014).

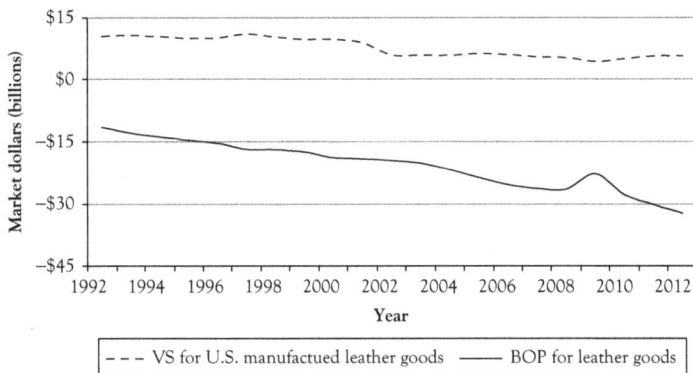

Figure 7.29 Comparison of the value of shipments of U.S.-manufactured leather goods with corresponding balance of payments

Source: U.S. Census Bureau (2013); ITA Trade with Selected Market (2014).

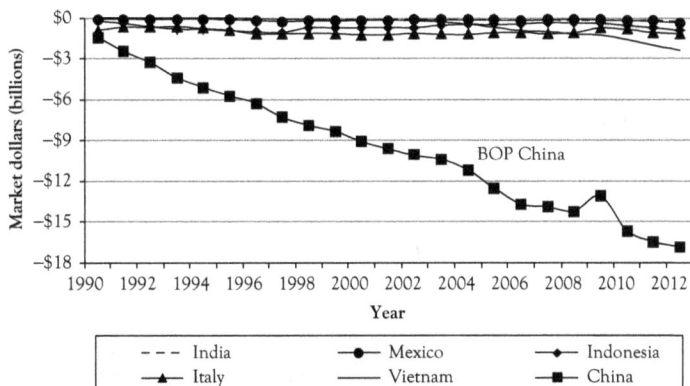

Figure 7.30 Balance of payments for footwear for the countries where the United States has the largest deficit

Source: ITA Trade with Selected Market (2014).

Did Anyone Get the Number of That Truck?

Manufacturing in the United States has been hit by a truck. Moreover, manufacturing industries in other countries have been hit by the same truck. The license plate on that truck is a vanity plate that reads "Made in China." Over the past 15 years, China has undergone a remarkable transformation and become a manufacturing juggernaut—seemingly unstoppable.

Figure 7.31 presents the U.S. trade deficits for those countries with which the United States has the largest deficit for durable goods. As may be seen, the deficit with China for durable goods is over $250 billion. The deficits with the other countries are all less than $100 billion. After China, the countries with which the United States has the largest trade deficits in durable goods are Japan, Mexico, Germany, and South Korea. For each of these countries, the industry for which the United States has the largest trade deficit is transportation equipment, particularly motor vehicles and parts. This is a manufacturing industry which China has not yet gone global, but motor vehicle parts industry is a rapidly growing export industry for China.

Figure 7.32 presents the U.S. trade deficits for those countries with which the United States has the largest deficit for nondurable goods. Once again, the deficit with China is much larger than the deficits for the other countries. The deficit with China is about $85 billion. The deficits with the other four countries are all under $20 billion. The deficit for nondurable goods with China arises primarily from apparel, leather goods, and textiles. Interestingly, the U.S. deficits with the other countries are again in areas where China has not yet gone global.

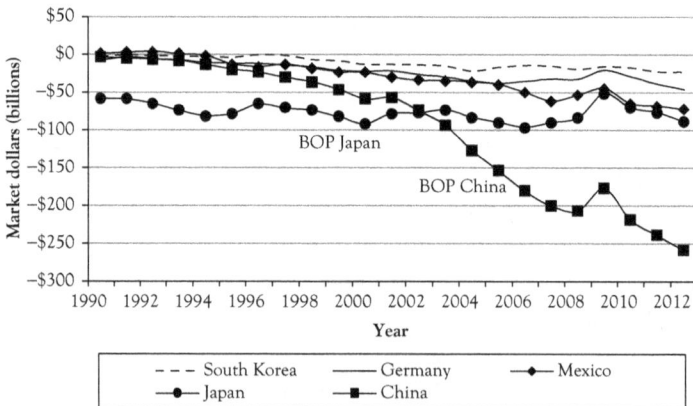

Figure 7.31 U.S. balance of payments trends for durable goods with those countries where the United States has the largest deficit in 2012

Source: ITA Trade with Selected Market (2014).

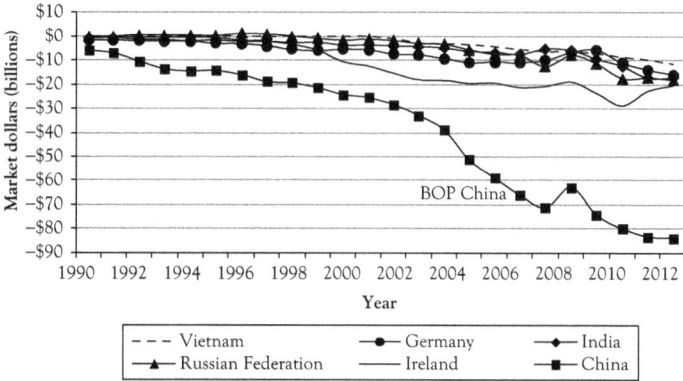

Figure 7.32 U.S. balance of payments for nondurable goods for those countries where the United States has the largest deficit

Source: ITA Trade with Selected Market (2014).

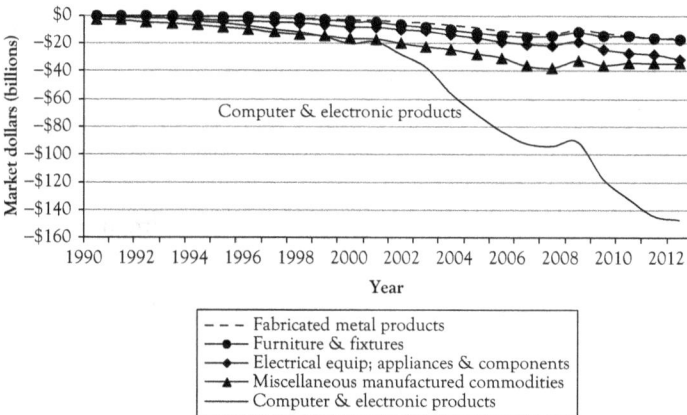

Figure 7.33 U.S. balance of payment trends with China for top five deficits in durable goods

Source: ITA Trade with Selected Market (2014).

For Ireland, India, and Germany, the nondurable U.S. deficits arise primarily from the importation of chemicals, many of which are likely to be used in pharmaceuticals. India also exports apparel to the United States as does Vietnam. The U.S. deficit in nondurable goods with the Russian Federation arises primarily from the importation of petroleum products.

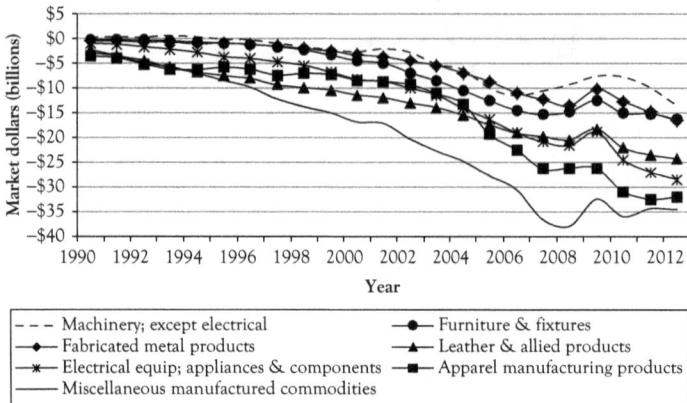

Figure 7.34 U.S. balance of payment trends with China for leading products other than computers

Source: ITA Trade with Selected Market (2014).

Figure 7.33 shows that the U.S. deficit with China for computers and electronic products at roughly $150 billion is over half of the total durable goods deficit with China. Computers and peripheral equipment, communication equipment, and audio and video equipment are dominating factors in the U.S. deficit with China. Figure 7.34 presents U.S. BOP trends for the leading industries other than computers and electronic products. The diversity of the industries is remarkable in terms of the breadth of penetration into the U.S. market. The rapid growth is also amazing. The truck is big and fast. The data in Figures 7.33 and 7.34 are summarized in Table 7.6 and in Figure 7.35.

Figure 7.36 presents the cost of U.S. imports of manufactured goods from China compared to the U.S. exports of manufactured goods to China. As is readily seen, imports to the United States have soared while exports to China have grown very slowly. In fact as seen in Figure 7.37, amazingly, the United States spends as much annually on items made in China (manufactured goods) as for all the oil and gas imported to the United States. It is hard to imagine or believe. Moreover, since the United States exports relatively little to China compared to the imports, the comparison of trade deficits is also hard to believe. Figure 7.38 shows that the deficit for oil is growing smaller so that the deficit for Chinese manufactured goods exceeds the deficit for oil by $74 billion. The data are summarized in Table 7.7.

Table 7.6 Industries for which the United States has the largest trade deficits with China (billions of current dollars)

	1990	1995	2000	2005	2010	2013
Computer & Electronic Products	–$0.90	–$5.53	–$20.45	–$70.83	–$117.54	–$147.06
Miscellaneous Manufactured Commodities	–$2.74	–$8.39	–$16.79	–$27.72	–$35.98	–$34.56
Apparel Manufacturing Products	–$3.46	–$5.74	–$8.31	–$19.30	–$30.94	–$32.82
Electrical Equipment; Appliances	–$0.83	–$3.55	–$8.16	–$16.33	–$24.42	–$31.37
Leather & Allied Products	–$2.15	–$7.43	–$11.32	–$17.35	–$22.04	–$24.27
Furniture & Fixtures	–$0.11	–$0.86	–$4.36	–$12.55	–$15.01	–$17.26
Fabricated Metal Products	–$0.33	–$0.89	–$3.21	–$8.62	–$12.58	–$16.52
Machinery; Except Electrical	$0.34	–$0.02	–$2.59	–$9.98	–$7.69	–$13.80
Plastics & Rubber Products	–$0.32	–$1.23	–$2.35	–$6.18	–$9.44	–$13.79
Textile Mills Products	–$0.46	–$1.09	–$1.74	–$5.67	–$8.18	–$9.41
Nonmetallic Mineral Products	–$0.19	–$0.89	–$2.17	–$3.87	–$4.57	–$5.59
Chemicals	$0.84	$1.35	$0.71	$0.82	–$0.70	–$2.66
Printed Matter & Related Products	–$0.04	–$0.24	–$0.62	–$1.61	–$2.11	–$2.20
Wood Products	–$0.14	–$0.34	–$0.75	–$2.00	–$2.16	–$2.14
Textiles & Fabrics	–$0.18	–$0.19	–$0.33	–$0.71	–$0.98	–$1.46
Primary Metal Mfg	–$0.09	–$0.26	–$0.74	–$2.35	–$0.48	–$1.28
Paper	$0.07	$0.01	–$0.05	–$0.86	–$0.63	–$0.79

Source: ITA Trade with Selected Market (2014).

Summary

It has been shown in this chapter that many U.S. manufacturing industries have experienced declining value of shipments while at the same time seeing growing imports take market share. These two trends generally mean a growing U.S. trade deficit in that industry. For a number of

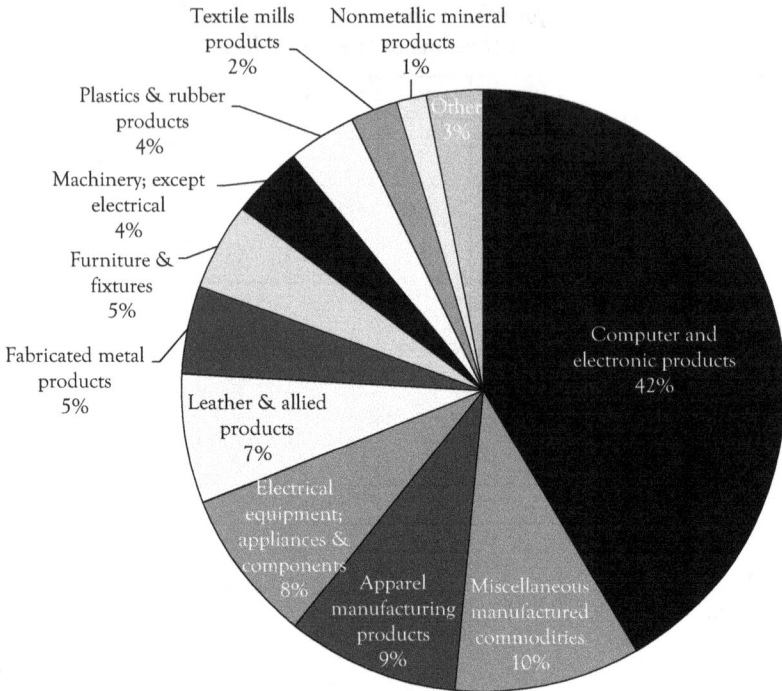

Figure 7.35 Distribution of U.S. imports from China
Source: ITA Trade with Selected Market (2014).

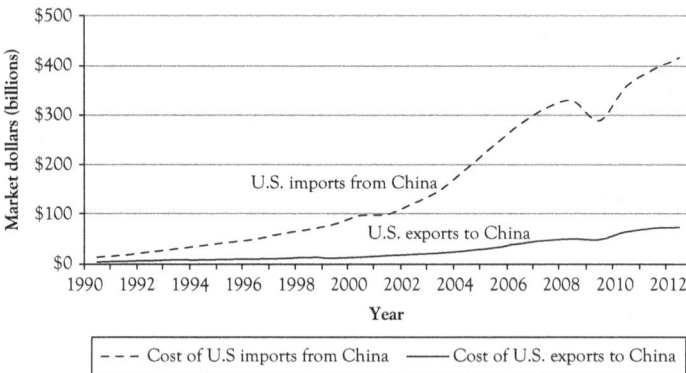

Figure 7.36 Costs of imported manufactured goods from China compared with United States export to China
Source: ITA Trade with Selected Market (2014).

these industries the source of the imports has been a rapidly growing penetration of U.S. markets by products made in China. These markets and products clearly include computers and peripheral products, communication equipment, apparel, and footwear. Moreover, China products appear

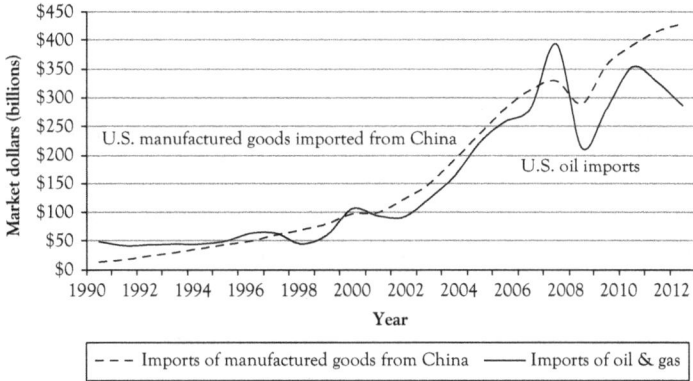

Figure 7.37 Cost of imported oil to the United States compared with the cost of manufactured goods imported from China

Source: ITA Trade with Selected Market (2014).

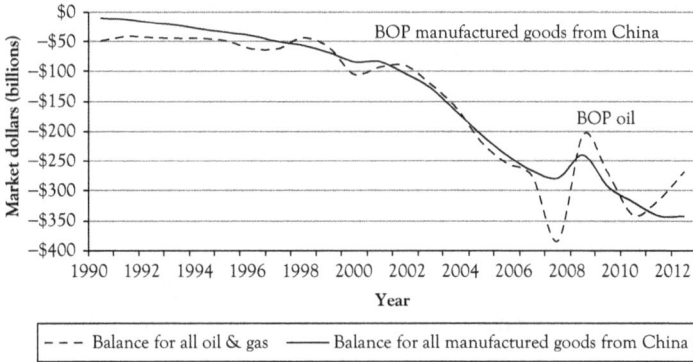

Figure 7.38 A comparison of the U.S. balance of payments for oil versus the balance of payments with China for manufactured goods

Source: ITA Trade with Selected Market (2014).

Table 7.7 U.S. balance of payments for oil and gas compared to balance of payments for manufactured goods from China (billions of current dollars)

	1990	1995	2000	2005	2010	2013
BOP for Oil & Gas Imports	−$48.30	−$49.13	−$104.61	−$217.90	−$270.55	−$268.88
BOP for Manufactured Goods from China	−$10.08	−$34.33	−$83.55	−$205.76	−$292.75	−$343.16

not only to displace products made in the United States but in other countries as well. The data, for example, clearly show falling U.S. imports of computers and peripheral products and audio and video products that are made in Japan. Both Malaysia and Taiwan have seen their computer and peripheral exports to the United States fall.

Mexico seems to be the only country that is holding on to shipments to the United States but is actually expanding certain product shipments particularly motor vehicles, motor vehicle parts, communication equipment, and audio and video equipment. NAFTA made "near-shoring" production to Mexico a very attractive proposition. As the Economic Policy Office put it just two years after NAFTA was enacted "Many imported goods, such as automobiles imported from Mexican assembly plants, replace goods that were made in U.S. factories that have closed or downsized" (Scott 1996).

Nevertheless, the United States spends more money for imports of manufactured goods from China than on imported oil and gas. As a result, the U.S. trade deficit for oil and gas is significantly smaller than the deficit for Chinese manufactured goods.

Key Takeaways

Three fundamental patterns have been seen in the figures in this chapter. Data Pattern 5 illustrates the stagnant or declining value of shipments for U.S.-manufactured products, growing imports, and a growing negative BOP for those same goods.

This pattern clearly indicates growing consumption on the part of U.S. consumers but with most of the growing demand being captured by imported products. This was most obviously true for computer and peripheral products, communication equipment, motor vehicle parts, apparel, and leather goods.

Data Pattern 6 illustrates rapidly growing imports from China and stagnant or declining imports from other countries. In a very real sense, the manufacturing industries in those other countries are being adversely impacted just as their counterparts in the United States. This was seen for example in the greatly reduced shipments of computers and TVs by Japan.

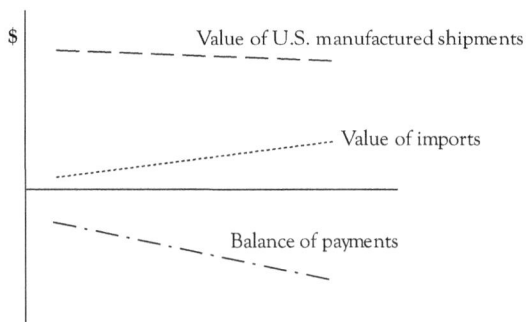

Data Pattern 5: Value of U.S.-manufactured shipments, imports, and balance of payments

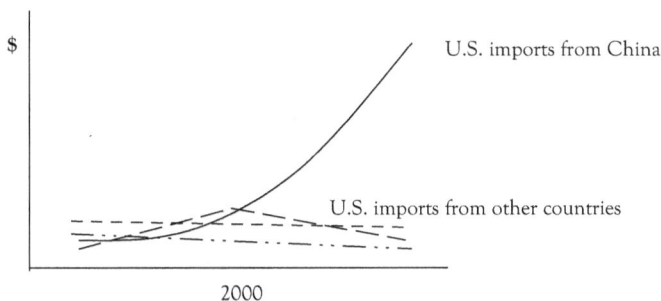

Data Pattern 6: Imports from China and imports from other countries

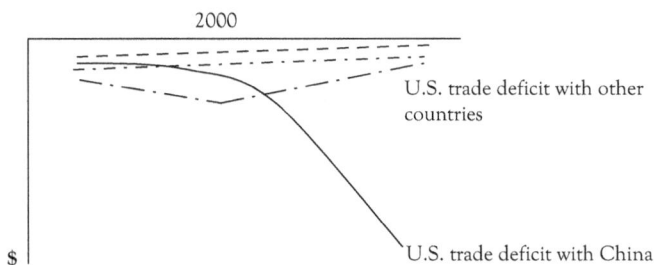

Data Pattern 7: Trade deficits with China and with other countries

Data Pattern 7 is essentially a reflection of the second pattern. This pattern includes a rapidly rising trade deficit with China and stagnant or declining deficits with other countries.

CHAPTER 8

Findings, a History of Warnings, and Corrective Policy Actions

Observations and Findings

It has been claimed by some that U.S. manufacturing is doing fine and that the loss of employment is the result of rapid growth in productivity. For example, "manufacturing employment has fallen (since 2000) because of productivity growth, not a decline in output" (Morrison and Labonte 2008). In Chapter 6, however, the exact opposite was shown for a significant number of manufacturing industries—output and value of shipments *have* declined and led to reductions in employment. If a company is not making the products, it does not need the workers. In what might be termed the farming analogy, manufacturing is simply following the path of agriculture in which ever greater output is achieved with fewer employees, see Figure 6.3a. For farming, the productivity growth occurred over decades with growing use of equipment, fertilizer, and pesticides, see Figure 6.5. At the same time, the average farm was growing in terms of acreage, see Figure 5.2. For U.S. manufacturing, however, the losses of employment and facilities have been over a devastatingly short period of time following the year 2000. And rather than a movement toward larger plants and factories, most manufacturing job losses were associated with the closing of large plants. And rather than growing output as in farming, many manufacturing industries have declining output. There are very few similarities between the long-term evolution of agriculture and the short-term losses of U.S. manufacturing. U.S. manufacturing is in a severe state of decline:

> Between 1999 and 2007, while the U.S. economy was growing, manufacturing lost 3.4 million jobs; another 2 million manufacturing

jobs were lost in the Great Recession between 2007 and 2012 (Bureau of Labor Statistics 2014a).

Between 1999 and 2007, the number of manufacturing plants and factories dropped by 50,000; another 25,000 establishments were lost during the recession between 2007 and 2012 (Bureau of Labor Statistics 2014a).

Our trade deficit for manufactured goods was $449 billion in 2013! (ITA Global Patterns of U.S. Trade 2014). This trade deficit for manufactured goods was $180 billion *in excess* of our $269 billion deficit for oil (ITA Global Patterns of U.S. Trade 2014).

Our trade deficit for manufactured goods *from China alone* was $343 billion in 2013, $74 billion dollars greater than our deficit for oil; that's worth repeating—our trade deficit for stuff imported from China is greater than our trade deficit for all the oil we import! (ITA Trade with Selected Market 2014).

The year 2000 was a watershed year for American manufacturing. Prior to that year, manufacturing employment had been stable or increasing in most industries with the exception of the apparel and aerospace industries. As discussed in Chapters 2 and 7, the U.S. apparel industry was impacted adversely by several events in the nineties: the signing of The North American Free Trade Agreement (NAFTA) in 1994, the expiring of the Multi-Fiber Agreement also in 1994, the Asian financial crisis in the late nineties which made their products less costly in the United States, Congress enacting legislation in 2000 granting China the status of Permanent Normal Trade Relations (PNTR), and, finally, China gaining admission to the WTO in 2001 (Clifford 2013). Each of these events contributed to greater importation of apparel. The aerospace industry was hit hard by the rapid decline of defense spending in the early nineties following the end of the Cold War. On the other hand, however, other U.S. manufacturing industries fared reasonably well between 1990 and 2000. Excluding aerospace and apparel, U.S. manufacturing industries added 311,000 jobs in the nineties. From 1990 to 2000, the number of manufacturing plants, facilities, and establishments in the United States grew from 387,000 to 405,000. Then around the year 2000, rapid deterioration set in. Between 2000 and 2007 (this is before the Great Recession, a seven-year period while the general economy was growing), U.S. manufacturing employment fell by 3.4 million jobs and the number of manufacturing establishments dropped by 50,000. Moreover, analysis of the establishment data including the number of employees shows that, over time, somewhat surprisingly, the vast majority of factory closings were the larger plants with the greatest number of employees. This was true for most industries—the largest factories and plants were the ones most likely to close.

Even with plunging employment and facilities, two important measures of total manufacturing output increased between 2000 and 2007, the real value added by the BEA and the industrial production (IP) index by the Federal Reserve. This would seem to be completely paradoxical. As was shown in Chapter 6, however, these government statistics distort the true situation because of the enormous growth in technical capabilities of computers that is part of estimating real value added and the IP index. This growth of technical capabilities in the products of one industry (computers and peripheral equipment) is so great that it creates an aura of health for all of manufacturing that is not reflected in terms of actual shipments. Atkinson computed corrective estimates and found that "When government measurement errors are corrected, it appears that real U.S. manufacturing output declined by 11 percent from 2000 to 2010, likely the only decade in American history (other than the Great Depression) where manufacturing output fell" (Atkinson et al. 2012). The fact that the value of shipments and production capacity are down for many manufacturing industries supports this position. Conditions are far from being as good as official statistics indicate. The loss of jobs is due to loss of production and output, not rapid growth in productivity. In truth, U.S. manufacturing is in trouble and that puts the economy and national security in jeopardy.

Seven data patterns have been identified in the previous chapters. Each of the seven patterns has been observed in numerous industries indicating a strong commonality of cause. The seven patterns are:

1. Stable or growing employment between 1990 and 1999, then a precipitous decline in employment;

2. Stable or growing number of establishments between 1990 and 1999, then a strong decline in the number of establishments;

3. Stable mix in the size of establishments between 1990 and 1999, then a sharp decline in the number of establishments with more than 500 employees and an even stronger decline in the number of establishments with more than 1,000 employees;

4. Growing IP, industrial capacity, and value of shipments between 1990 and 1999, then stagnation or decline in these three metrics of manufacturing industry health;

5. Beginning in 2000, stagnation or decline in the value of U.S.-manufactured shipments, strong growth in imports, and strong growth in the trade deficit;

6. Beginning in 2000, very strong or explosive growth in imports from China and stagnation, or more likely, declines in imports from other countries; and

7. Beginning in 2000, rapid growth in the trade deficit with China.

These patterns of stagnation and decline were observed across many industries. It would be very difficult, if not impossible, to argue plausibly that productivity growth hit many manufacturing industries simultaneously beginning in 2000 to cause the loss of employment and establishments. Especially since different industries achieve productivity gains at different speeds and with different approaches. Moreover, productivity growth generally requires capital investment for machinery and equipment. As Atkinson has shown, however, "Most U.S. manufacturing industries now have less machinery and equipment than they did a decade ago..." (Atkinson et al. 2012). It would seem clear at this point that off-shoring and trade with China, and, to a lesser extent, with Mexico are behind the massive loss of jobs and manufacturing establishments that began after 1999 and 2000.

A History of Warnings

1985

Berkshire Hathaway announced it would close its last textile plant located in New Bedford, Mass. and withdraw from that business (New York Times

1985). The president of the division said the closing "was a response to increased competition from Japan and Taiwan, depressed prices and the strength of the dollar."

1990

Canada, the United States, and Mexico agree to pursue a free trade agreement that will ultimately become NAFTA.

1992 and 1993

President George H. W. Bush and the presidents of Mexico and Canada sign the NAFTA. The signed agreement then had to be approved by each nation's legislative branch. As the debate raged in the United States, Ross Perot warned that NAFTA would cause American industry and agriculture to flee south with "a giant sucking sound" toward cheaper Mexican labor (The New York Times 1993). On the other side of the debate, Lee A. Iacocca, then the recently retired chairman of Chrysler, appeared in pro-NAFTA TV advertisements citing job creation for many American workers (Dale 1993). Big business was strongly behind NAFTA and their lobbying campaign was "running wide open, producing mailings, telephone calls, newspaper advertisements and speeches on behalf of the NAFTA. Hundreds of business executives are spending more time on Capitol Hill than at home" (New York Times 1993). Labor was strongly opposed to NAFTA fearing the loss of jobs to Mexico. The American Federation of Labor-Congress of Industrial Organizations spent more than $3 million for its nationwide campaign, and that was supplemented by funding from the machinists, teamsters, auto workers, and garment industry unions, and the unions were joined by environmental, religious, civil rights, farming, and immigrant rights groups (Chicago Tribune 1993). NAFTA was passed by the House of Representatives (234 to 200) and the Senate (61–38) in late 1993. President Clinton signed it into law in the December of 1993 and it became effective on January 1, 1994.

The battle for NAFTA foreshadowed the China PNTR and WTO debate with big business supporting the trade pact, and workers, environmentalists, and human rights groups opposing.

In January 1992 the United States government signed a Market Access Memorandum of Understanding (MOU) with China in which China agreed to lower tariffs, eliminate import-substitution requirements, and reduce other regulatory impediments to imports. Also signed was an Intellectual Property MOU in which China agreed to change its copyright and patent laws to substantially improve protection of foreign literary works, recordings, computer software, manufacturing processes, and product design. Again, the MOUs were hailed as offering incredible opportunities for export of U.S. goods to China and for job creation. In 1997, Peter Morici, the former director of the Office of Economics in the U.S. International Trade Commission (USITC) wrote "These MOU's ... have not created all the exports expected, in part, because China has not fully abided by the terms of these agreements" (Morici 1997). As a result, the U.S. trade deficit with China continued to grow after signing of the MOUs. Morici also noted regarding the deficit:

> Efforts to correct this imbalance have been frustrated by the tendency of Chinese officials to replace import barriers eliminated under the Market Access MOU with new ones. For example, quotas on imports of medical equipment and film have been replaced by complex registration and tendering requirements. ... Further, Chinese policymakers are imposing tough local sourcing and technology transfer requirements on foreign investors; these requirements reduce Chinese imports of American products and their effect will grow over time.

1994–1996

Reports about the negative impacts of U.S. trade policies began within a few years after NAFTA was enacted on January 1, 1994. Just two years later, the Economic Policy Institute (EPI) wrote "Our best estimate, then, is that between 392,000 and 484,000 jobs were lost between 1993 and 1995 because of increasing trade deficits with the NAFTA countries" (Scott 1996). After NAFTA negotiations were announced in 1990, U.S. investment in new factories to make products for export to the United States rapidly accelerated, with many plants being built in the maquiladora zone along the U.S. border. As EPI noted "There were approximately 2,000 such plants in 1994. News reports in early 1995 said that Mexican authorities were approving applications for two to three new plants per day in this zone, and that 250 plant applications had been approved in the first quarter alone" (Scott 1996).

1997

In 1997, EPI reported that NAFTA had caused job losses in every state but that "Several states, notably Alabama, Arkansas, Indiana, Michigan, North Carolina, Tennessee, and Texas, experienced job losses disproportionate to their share of the overall U.S. labor force. These states have high concentrations of industries (such as motor vehicles, textiles, apparel, computers, and electrical appliances in which a significant amount of production has moved to Mexico) (Rothstein and Scott September 1997). Other researchers found that NAFTA had impacts beyond the workers who were actually displaced. In a report that was commissioned by the tri-national Labor Secretariat of the Commission for Labor Cooperation (NAFTA's own labor commission), it was found that the factory shutdown rate within two years of union certification was triple the rate before NAFTA went into effect (Bronfenbrenner 1996). The study also found that by threatening to close and move production to Mexico, many companies were able to win wage and benefit concessions from their workers (Bronfenbrenner 1997).

Warnings about China continued. A 1997 study estimated job losses for manufacturing industries with major trade deficits with China: apparel (146,000 jobs lost), toys (66,000 jobs lost), footwear (60,000 jobs lost), textiles (58,000 jobs lost), and consumer electronic devices (29,000 jobs lost) (Rothstein and Scott October 1997). This report also noted that China was moving up the product ladder from apparel, footwear, and toys to computers and electronics, and that job losses among higher wage workers would grow. This prediction certainly came true as the trade deficit soared after 1997. A Commission formed by President Clinton in 1995 identified several major factors driving the U.S. trade deficit with China and the Asia Pacific region (Commission on U.S.–Pacific Trade and Investment Policy 1997). These barriers that inhibited U.S. exports included non-tariff barriers such as licensing of production, piracy of technology and intellectual property, and undervalued currency due to intervention. The Commission noted:

> U.S. exports and investment are impaired by the persistence of private and public anti-competitive practices and poor enforcement of antitrust laws. Many Asia Pacific governments actively intervene in their domestic markets to protect favored industries. Furthermore, government bureaucracies in some countries use administrative guidance to intervene in the economy to influence business decisions to bolster exports and limit foreign direct investment. Many Asia-Pacific economies keep foreign firms out of the domestic market through preferential business relationships, price-fixing, bid-rigging, and market allocation arrangements among competitors.

This hardly sounds like "free trade" but U.S. leaders generally did not take action, or if they did, they backpedaled. As Morici notes:

> China frequently practices brinkmanship when disputes erupt, and then tests United States tolerance for violations in the resulting agreements. In part, this is attributable to the communist legacy, a weak legal system, and a Chinese tradition of bending rules to accommodate powerful interests. However, the United States has encouraged China to test its resolve by not responding to Chinese transgressions consistently. For example, import-substitution requirements imposed by China's 1994 automobile industrial policy violate the 1992 Market Access MOU and could result in more lost American jobs than CD piracy; yet, these requirements have not attracted United States threats of massive retaliation against Chinese apparel exports. Similarly, by back

peddling on trade sanctions related to high-policy issues, such as human rights and arms proliferation, United States policymakers encourage Chinese leaders and their conservative critics to question American resolve (Morici 1997).

1999

In 1993, the United States had a net export trade deficit with its NAFTA partners of $18.2 billion. By 1998, in just a five-year period, this deficit had soared by 160 percent to $47.3 billion (all figures in 1987 inflation-adjusted dollars) (Scott 1999). This report estimated that between 1994 and 1998, the growing trade deficit with Mexico and Canada had destroyed 440,172 jobs.

In 1998, the U.S. Trade Representative (USTR) asked the USITC to prepare a report assessing the probable economic effects on the United States of China's accession to the WTO. The USTR requested that the USITC use formal economic analysis to provide, to the extent possible, a quantitative assessment of the effects on the U.S. economy of China's WTO membership. The Commission's report "Assessment of the Economic Effects on the United States of China's Accession to the WTO" was released in 1999 (U.S. International Trade Commission 1999). In the report, the Commission used an analytical model that estimated U.S. exports to China would grow at 10.1 percent and that U.S. imports from China would grow at 6.9 percent. However, since imports in 1999 were $81 billion and exports were only $13 billion, these two growth percentages yielded a forecast of continuing growing deficit as shown in Table 8.1.

Even though these growing deficits implied continuing job losses in the United States, President Clinton continued to push for passage of PNTR for China and for China's membership in the WTO (Scott 2000). As distressing as the Commission's forecast appears, reality has actually been worse. Table 8.2 compares the Commission forecast for the deficit with the actual deficit. As may be seen, the actual deficit grew much faster than the forecast and by 2013 was actually twice the Commission's negative forecast.

Table 8.1 1999 forecast of U.S. imports and exports with China with USITC growth rates

Year	Forecast U.S. Imports from China	Forecast U.S. Exports to China	Forecast BOP with China
1999	$81.00	$13.00	($68.00)
2000	$86.59	$14.31	($72.28)
2001	$92.56	$15.76	($76.81)
2002	$98.95	$17.35	($81.60)
2003	$105.78	$19.10	($86.68)
2004	$113.08	$21.03	($92.04)
2005	$120.88	$23.16	($97.72)
2006	$129.22	$25.49	($103.72)
2007	$138.14	$28.07	($110.07)
2008	$147.67	$30.91	($116.76)
2009	$157.86	$34.03	($123.83)
2010	$168.75	$37.46	($131.29)
2011	$180.39	$41.25	($139.15)
2012	$192.84	$45.41	($147.43)
2013	$206.15	$50.00	($156.15)

Table 8.2 Comparison between actual and forecast U.S. balance of payments with China

Year	Actual U.S. BOP with China	USITC Forecast BOP with China
1999	–$68.68	–$68.00
2000	–$83.83	–$72.28
2001	–$83.10	–$76.81
2002	–$103.06	–$81.60
2003	–$124.07	–$86.68
2004	–$162.25	–$92.04
2005	–$202.28	–$97.72
2006	–$234.10	–$103.72
2007	–$258.51	–$110.07
2008	–$268.04	–$116.76
2009	–$226.88	–$123.83
2010	–$273.04	–$131.29
2011	–$295.39	–$139.15
2012	–$315.10	–$147.43
2013	–$318.42	–$156.15

2000

In the year 2000, the United States granted PNTR to China which paved the way for China to become a member of WTO. Prior to this, China was subject to annual reviews that were uncertain and politically contentious. Many companies, if not most, were hesitant to invest in China or to move production to China because of that uncertainty regarding tariffs. The debate in 2000 over PNTR was highly charged regarding the possible implications that the China trade bill might have for the U.S. economy, particularly the impact on wages and employment for U.S. workers. As with NAFTA, there was big business on one side actively lobbying for PNTR. The National Association of Manufacturers' President Jerry Jasinowski argued that, across the board PNTR would "increase opportunities for American manufacturers and their workers" (Bronfenbrenner et al. 2001). As described by Bronfenbrenner,

> Those opposing PNTR presented a very different view on its possible effects on the U.S. economy. Ohio Republican, Representative Bob Ney warned of the impact of the bill on American jobs. "This is a greed agreement. . . . It only helps a few at the top," said Rep. Ney. "We're going to lose hundreds of thousands of jobs here in America." His words were echoed by then United Steelworkers of America President George Becker who called the drive for PNTR "nothing more than greedy multinationals capitalizing on Chinese labor at dirt-cheap prices."

On May 24, 2000, Representative Richard Gephardt made the following remarks in opposition to PNTR with China (Congressional Record—House 2000):

> Some would argue that this is just about trade. I would remind them that our greatest export is not our products and our services, our greatest exports are our ideals and our values. Getting acceptance of these ideals is also vital for trade. A country that fails to respect basic rights of people will not respect the rule of law, and with rule of law in China, the rights of businesses will not be accepted.

China has not obeyed the agreements that have been made with us on trade. We have been promised access; we have not gotten it. We have been promised protection of intellectual property; we have not gotten it. Our trade deficit is now $85 billion with China, the highest as a percent of total trade of any country in the world. We export more now to Singapore, a nation of 3.5 million people, than we export to China, a country of 1.3 billion people. The track record is poor on compliance with treaties. Let us not reward them before we get them to comply. China's leaders show contempt for the rule of law.

The House of Representatives passed the bill granting PNTR to China on May 24, 2000 by a vote of 237 to 197 (International Center for Trade and Sustainable Development 2000). The Senate passed the legislation on September 19, 2000. Following the Senate vote, President Clinton made the following remarks at a press conference in the White House:

> When we open markets abroad to U.S. goods, we open opportunities at home. This vote will do that. In return for normal trade relations—the same terms of trade we offer now to more than 130 other countries—China will open its markets to American products from wheat, to cars, to consulting services. And we will be far more able to sell goods in China without moving our factories there (Clinton 2000).

Unfortunately, this rosy outlook did not come to pass.

2001

On December 11, 2001, China became a member of the WTO.

In the fall of 2000, legislation was enacted by the U.S. Congress to establish a bipartisan commission to investigate, assess, and report to Congress on the economic and security implications of the bilateral economic relationship between the United States and China. Because no U.S. government agency had the responsibility for collecting data on the wage and employment effects of trade agreements and policies, the U.S. Trade

Deficit Review Commission initiated a pilot study with two tasks. The first task was to monitor and analyze media coverage of the employment and wage effects of China trade and investment by tracking all media-reported production shifts out of the United States to China, Mexico, and other Asian and Latin American countries and out of Asian and Latin American countries into China that occurred between October 1, 2000 and April 30, 2001. The second task of the study involved collecting and analyzing macro data on imports, exports, and foreign direct investment in those industries and economic sectors that have an active trade, investment, and production relationship with China (Bronfenbrenner et al. 2001).

The findings of the study were dramatic.

In the months since the enactment of PNTR legislation with China there has been an escalation of production shifts out of the U.S. and into China. According to our media-tracking data, between October 1, 2000 and April 30, 2001 more than eighty corporations announced their intentions to shift production to China, with the number of announced production shifts increasing each month from two per month in October to November to nineteen per month by April. The estimated number of jobs lost through these production shifts to China was as high as 34,900, compared to 29,267 jobs lost to Mexico, 9,061 jobs lost to other Asian countries, and fewer than 1,000 jobs lost to other Latin American countries. However, because we believe our media tracking captures fewer than half of all production shifts out of the U.S. to China and other countries during this period, we estimate that the actual number of jobs lost through production shifts to China and Mexico averages between 70,000 and 100,000 jobs each year for each country. This is in keeping with our preliminary macroeconomic analysis of the employment affects of U.S.-China trade balance that estimates as many 760,000 U.S. jobs have been lost due to the U.S.-China trade deficit since 1992 (Bronfenbrenner et al. 2001).

The study found that the companies shutting down and moving to China were generally large, profitable, and well-established. This result agrees with the finding in this analysis that most plant closings were in the two largest size groupings. Some of the companies identified in the study moving to China included well-known names such as Lexmark (printers), Mattel Murray (Barbie doll playhouses), Raleigh (bicycles), Motorola (cell phones), and Samsonite (luggage).

A strong statement summarized the studies observations: "In conclusion, our research suggests that the U.S. and other countries have moved ahead with trade policies and global economic integration based on faulty arguments and incomplete information" (Bronfenbrenner et al. 2001). The implications of this are, of course, millions of jobs lost in the United States, entire communities being ruined, depressed wages for workers not displaced, and increasing income inequality.

2004

Even with mounting evidence that trade was having a very large negative impact on U.S. manufacturing, some reports continued to point to shifting consumer demands and productivity gains as drivers of manufacturing's decline. A 2004 report from the Congressional Budget Office (CBO) (Congressional Budget Office 2004) stated that labor productivity in manufacturing has risen at an average annual rate of 5.5 percent, faster than its average annual rate of growth during previous postwar recessions and the early part of the ensuing recoveries. As shown, however, in Chapter 6, productivity for manufacturing is badly overstated due to the treatment of computers and peripheral equipment. This is a completely misleading argument. The CBO paper also argued "the growth in demand for manufactured goods has not kept pace with the growth in productivity, as consumers continue to devote more of their spending to services instead of goods."

On-going reports on the effects of the rapidly rising U.S. trade deficit, however, continued to counter the productivity and shifting demand arguments such as in the CBO paper. The EPI report "Shifting Blame for Manufacturing Job Loss" had the following conclusions:

U.S. consumers and businesses have not shifted their purchasing away from manufactured goods. In fact, demand for manufactured goods as a share of total demand in the United States has actually grown over the past 10 years.

The rising trade deficit has led to an unprecedented divergence between domestic manufacturing output and demand. Domestic output is now just 76.5 percent of domestic demand, nearly 14 percent less than the 1987 to 1997 average. Raising output closer to this previous relationship with demand (around 90 percent) would generate millions of jobs in manufacturing.

The rising trade deficit in manufactured goods accounts for about 58 percent of the decline in manufacturing employment between 1998 and 2003 and 34 percent of the decline from 2000 to 2003. This translates into about 1.78 million jobs since 1998 and 935,000 jobs since 2000 that have been lost due to rising net manufactured imports (Bevins 2004).

2006

The entry of China, India, and the ex-Soviet bloc countries into the global economy greatly increased the size of the global labor pool—from roughly 1.5 billion workers to 3 billion workers. Richard Freeman called this "The Great Doubling" (Freeman 2006). As Freeman notes, this has shifted the global balance of power to capital and away from labor. With this huge new supply of labor, firms can off-shore their production to lower wage countries or threaten to do so unless workers grant concessions in wages and benefits. Retailers can import products and manufacturers can import parts from lower wage countries. According to Freeman, the doubling of the global workforce creates two major threats to worker well-being in the United States and other advanced countries. "First, it creates downward pressures on the employment and earnings of less skilled workers through trade and immigration." The second "is that these countries are becoming competitive in technologically advanced activities." As Freeman notes, between 1999 and 2005, China increased the number of bachelor degrees awarded fivefold to four million graduates. The combination of low wages and well educated workers makes these countries intimidating competitors for any advanced country and threaten workers at all skill levels.

2007

A 2007 EPI report (Scott 2007) found that the rise in the U.S. trade deficit with China between 1997 and 2006 had displaced U.S. production that would have supported 2,166,000 U.S. jobs. The report noted that most of the jobs, 1.8 million, had been lost since China joined the WTO in 2001. The report noted four key reasons for the rapid growth in imports from China: (i) currency intervention to keep the yuan artificially undervalued; (ii) suppression of labor rights to keep wages low; (iii) large scale direct subsidization of export production; and (iv) strict non-tariff barriers to imports. The report concluded: "Simply put, the promised benefits of trade liberalization with China have been unfulfilled."

Ferguson and Schularick (Ferguson and Schularick 2007) noted that the integration of China into the world economy doubled the global labor force which then boosted returns on capital and corporate profitability. As may be seen in Figure 8.1, U.S. corporate after-tax profits soared between 2001 and 2006, nearly tripling from $500 billion to $1.4 trillion.

Figure 8.2 presents both the U.S. corporate after-tax profits and U.S. imports from China. The similarity in timing of the trends is remarkable. As had been warned earlier, corporations and their shareholders benefited from lower-cost Chinese workers and American workers suffered.

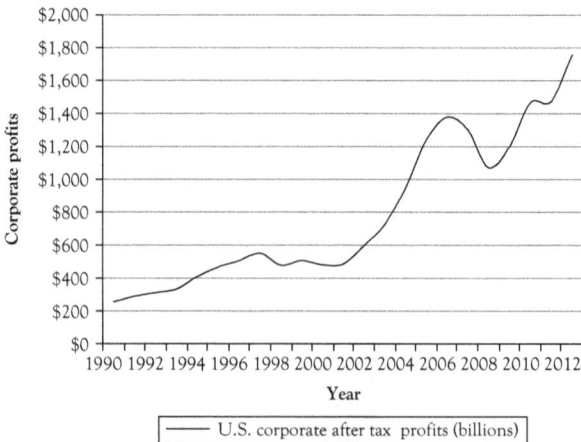

Figure 8.1 U.S. corporate after-tax profits

Ferguson (Ferguson and Schularick 2007) noted, however, that the higher returns on capital had not led to a higher cost of capital. As they argued, in a "standard neoclassical model, the rate of return on capital is equal to the marginal productivity of capital which, under perfect conditions, equals the cost of capital." However, during this period, global real interest rates did not reflect the increase in the return on capital that has taken place as a result of globalization, in fact, interest rates remained low. Given that the export performance of Chinese companies was responsible for the massive increase in the Chinese current account surplus, Ferguson argued that the root cause of the Chinese profits to be the exchange rate and its impact on the competitiveness of Chinese production. In other words, the commitment of Asian governments to fixed exchange rates contributed to high savings by artificially stimulating exports and decreasing import demand. Chinese companies—many of them state owned—have taken over large parts of the domestic market from foreign competition, depressing imports and expanded their market share abroad, increasing exports. Profits have surged and the dollars have piled up at the People's Bank of China with the export of excess savings to the United States, depressing United States and global interest rates. As others before them, Ferguson and Schularick argue that the undervalued renminbi has played a key role in the loss of American jobs.

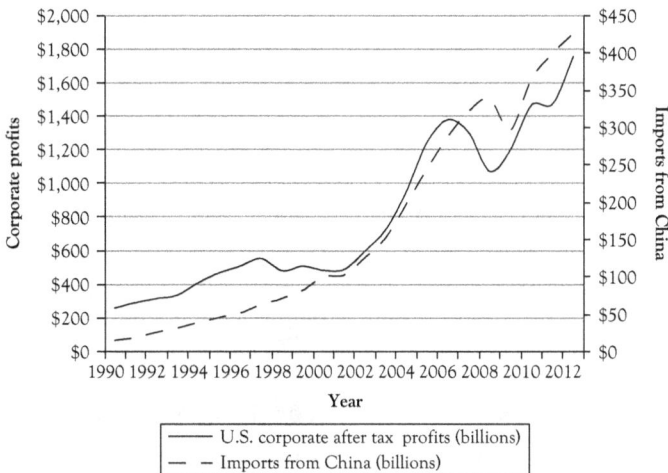

Figure 8.2 U.S. coporate after-tax profits and imports from China

2009

In 2009, Ferguson and Schularick updated their analysis of 2007. In the 2009 paper (Ferguson and Schularick 2009), they point out that the Chinese economy underwent rapid gains in productivity over the past decade. As a result, unit labor costs continued to fall for most of the period in absolute terms and relative to other countries. These gains, however, were not translated into exchange rate realignments. This led to massive gains in competitiveness for China. Ferguson and Schularick "think there is strong evidence that productivity-adjusted production in China today is 40 percent cheaper than a decade ago." As a result, trade surpluses have surged and corporate profits of Chinese firms have soared, resulting in huge reserve accounts. In their opinion, currency adjustment is much needed. Their calculations indicate that the current exchange rate, after adjusting for differences in productivity, is undervalued by somewhere between 30 and 48 percent. The consequences of this exchange undervaluation have become too big for the world economy to bear. Simply put, because Beijing keeps the exchange rate fixed, the dollar cannot devalue against China (and other parts of Asia) despite the large U.S. trade deficits. This makes it impossible for the American economy to earn its way out of the slump. Without an exchange rate adjustment, the United States will be forced to run expansive domestic policies if it wants to achieve full employment but deficit concerns make those policies very difficult to implement.

Cline and Williamson (Cline and Williamson 2009) calculated that the three countries with the largest undervaluations were China (with a needed appreciation of 21.4 percent), Malaysia (18 percent), and Taiwan (13.8 percent). The largest overvaluations were estimated to be those of the United States (with a needed depreciation of 17.4 percent), South Africa (13.2 percent), and Australia (11.9 percent). This condition makes U.S. products overly costly than products made in China, Malaysia, and Taiwan. As noted by Ferguson (Ferguson and Schularick 2009), this creates the situation in which the United States cannot earn its way out. Cline concludes by noting that "it is important that China changes its peg from the dollar to a basket to stabilize the effective rate."

Webber (Webber 2009) conducted a study to examine the erosion of industrial capabilities, the impact on U.S. defense, and the growing dependency on foreign goods to fill this gap. Webber evaluated the health of 16 manufacturing industries that are within the defense industrial base and are key to innovation and to the production of innovative products. Webber used three indicators as a measure of industry health: employment, shipments, and the number of establishments. These are metrics similar to the ones presented in Chapters 2, 3, and 6 of this book. Webber found that 13 of the 16 industries exhibited a high level of erosion. In general, Webber's results agree with and support the analysis presented in this book. As shown in Chapter 7, Navigational and Electronic Instruments was the only industry in the Computer and Electronic Products subsector that had growing value of shipments and a small positive trade balance. However, most of the other industries were shown to have declining employment, reduced number of establishments, and stagnant or declining sales. Table 8.3 presents a summary of Webber's results and the strong erosion that has taken place in many industries critical to our national defense.

2010

Webber's work regarding the defense industry was continued by Yudken in the report "Manufacturing Insecurity" (Yudken 2010). Yudken focused primarily on the substantial declines in the semiconductor, printed circuit board, machine tools, advanced materials, and aerospace industries. For semiconductors, Yudken reported that nearly one-half of the U.S. semiconductor market was supplied by imports in 2007. The printed circuit board industry has nearly disappeared in the United States "The U.S. PCB industry once dominated global PCB production, with 42 percent of global revenues in 1984, falling to 30 percent in 1998 and to less than 8 percent in 2008." By 2007, China or Hong Kong had the largest global market share accounting for 28 percent of worldwide PCB production. Machine tools, the foundation of the manufacturing process for products using metal, are critical in cutting and forming metals. Yudken paints a picture of severe erosion in this industry. "U.S. machine tool shipments fell to $2.2 million in 2003, the lowest level, in constant dollars, since

Table 8.3 Erosion of selected defense industrial support base sectors

		Employment	Economic Activity	Establishments	Overall Status
3315	Foundries	●	⊘	●	Eroded
33211	Forging & Stamping	●	●	●	Eroded
33271	Machine Shops	○	⊘	⊘	Healthy
332811	Metal Heat Treating	⊘	●	●	Eroded
332997	Industrial Pattern Manufacturing	●	●	●	Eroded
333295	Semiconductor Machinery	⊘	⊘	⊘	Holding Steady
333314	Optical Instrument and Lens	⊘	●	●	Eroded
333511	Industrial Mold Manufacturing	●	●	●	Eroded
333512	Machine Tools (Metal Cutting)	●	●	●	Eroded
333513	Machine Tools (Metal Forming)	●	●	●	Eroded
333514	Special Die & Tool, Die Set, Jig	●	●	●	Eroded
334412	Bare Printed Circuit Boards	●	●	●	Eroded
334413	Semiconductor & Related Devices	●	●	⊘	Eroded
334418	Printed Circuit Assemblies	●	●	○	Eroded
3345	Nav. Meas. & Control Instruments	●	○	⊘	Healthy
33591	Battery Manufacturing	●	●	●	Eroded

● Indicator eroded   ○ Indicator expanded   ⊘ Indicator held steady or showed signs of recovery

Source: Michael Webber, "Erosion of the U.S. Defense Industrial Support Base." In Richard McCormack (ed.), *Manufacturing A Better Future for America*, Washington, DC: Alliance for American Manufacturing (2009), 245–280: 274, Figure 3.

industry data began to be tracked in the 1920s." Yudken noted that between 1997 and 2007, the import penetration rate for metal-forming machine tools rose from 63 to 91 percent. The United States is rapidly losing these critical industries and, also, the potential for manufacturing products for national security.

An EPI report (Scott 2010) estimates that between 2001, with China's entry into the WTO, and 2008, 2.4 million jobs have been lost in the United States due to the rapid growth in the U.S. trade deficit with China. The report also notes that these job losses have also driven down wages and bargaining power for other workers in manufacturing and related industries.

2013

The U.S. Business and Industry Council published a report (Tonelson 2013, January, April) in which their analysis indicated "that, contrary to widespread optimism about an American industrial renaissance, domestic manufacturing's highest value sectors keep falling behind foreign-based rivals." The report found that 98 of 106 advanced manufacturing industries lost market share in their home U.S. market between 1997 and 2011. These industries losing home market share included a "long list of America's economic and technological crown jewels, including semiconductors; electro-medical apparatus; pharmaceuticals; turbines and turbine generator sets; construction equipment; farm machinery and equipment; mining machinery and equipment; several machine tool-related categories; and ball and roller bearings." The report included three findings with direct implications for policy. First, the latest data expose as "wishful thinking the claims made by President Obama, the Boston Consulting Group, and many other prominent voices that U.S.-based manufacturing has embarked on or is poised to launch an epic competitive rebound." Critical industries continue to lose market share. Figures in Chapter 7 showed the strong downward trends with ever growing deficits. These long-term trends are hard to reverse with continuing shrinkage and loss of market share by U.S. industries. Secondly, the import penetration data show "the conventional economic stimulus strategies still vying for supremacy in Washington—especially tax cutting to promote more capital spending—remain likely to produce increasingly disappointing results." The argument presented in the report is that most of the critical industries losing market share are capital goods industries that have been losing market share, if not shrinking, so that policies that impact these industries contribute less and less to the economy. Thirdly, the rapidly growing import penetration rates keep "undermining the rationale behind President Obama's decision to limit his trade-related economic recovery policies overwhelmingly to export promotion." Policies that generate significant "growth and employment multipliers will require U.S. policies aimed expressly at reducing these import penetration rates—i.e., at substituting domestically produced goods for products that Americans currently purchase from abroad." This report concludes that policies must have "the imperative of limiting—and indeed reducing—imports in order to spur manufacturing and overall economic growth."

A follow-on report from the U.S. Business and Industry Council (Tonelson 2013, April) showed that Chinese penetration in advanced U.S. manufacturing markets overall proceeded very rapidly between 1997 and 2011. During this period, for example, "China's share of American consumption of miscellaneous commercial and service industry machinery has skyrocketed by more than 36,000 percent. Its share of America's market for broadcast and wireless communications equipment has soared by 3,547 percent. The share of commercial heating and cooling equipment supplied to Americans by China is up 1,979 percent. China's market share in speed changers, high speed drives, and gears has jumped 2,131 percent. And in 18 of the other advanced manufacturing categories studied, Chinese import penetration rates rose by more than 1,000 percent between 1997 and 2011." These gains have moved China ahead of Germany and Japan as the largest sources of high-valued manufactured products. Supporting the rising deficit trends seen in Chapter 7 of this book, the U.S. Business and Industry Council report notes that "despite the proliferating claims of improved American manufacturing competitiveness, it seems clear that much more U.S. demand for these sophisticated products was met by these foreign producers than from producers in the United States." This is exactly the phenomenon seen for many manufacturing industries in Chapters 6 and 7—declining or stagnant value of shipments accompanied by rising imports and rapidly rising deficits.

MIT economics professor David Autor (Autor et al. 2013) found that rising imports led to higher unemployment, lower labor force participation, and reduced wages in local labor markets that contained manufacturing industries that compete with imports. The reduced wages were shown to extend to non-manufacturing sectors in the impacted markets. The study also found that payments for unemployment, disability, retirement, and healthcare also rose sharply in the more trade-exposed labor markets.

At this point it seems very clear that free trade has reduced U.S. manufacturing employment and depressed wages for those workers who were not displaced. Moreover, many of the workers who lost their manufacturing jobs were forced to take jobs at lower wages. Bevins (Bevins 2013) using a trade model estimates that the impact of trade on U.S. wages has been to reduce by 5.5 percent the average wage of the full-time, full-year worker earning average wage for workers without a four-year college degree. This echoes the remarks made in 2008 by Nobel Prize winning economist Paul Krugman who noted "that there has been a dramatic increase in manufactured imports from developing countries since the early 1990s. And it is probably true that this increase has been a force for greater inequality in the United States and other developed countries."

Figure 8.1 shows the strong growth in after-tax corporate profits in terms of dollars. One possible explanation for the growing profits would simply be that total profits increase with growth in the economy. That certainly has to be part of the story, but even more revealing is the growth of corporate profits as a percent of GDP as shown in Figure 8.3. This figure shows that the corporate profits as a percentage of GDP have doubled since the year 2000—growing from 4.6 to 10.8 percent. On the other hand, as shown in Figure 8.4, total wages as a percent of GDP have fallen from 47 to 42.6 percent since 2000.

Growing corporate profits and declining wages, expressed as percentages of GDP, would point to a growing inequality in incomes given that higher income groups are more likely to own the equity of corporations. As shown in Figure 8.5, beginning in the early 1980s, the share of U.S. income going to the top 1 percent began to grow after decades of stability (Saez 2013). With globalization taking off in the late 80s and

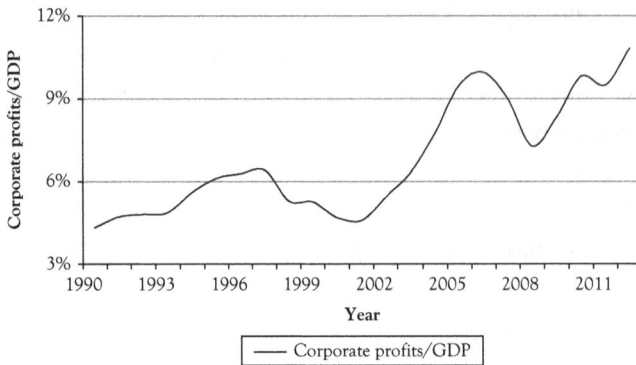

Figure 8.3 *Corporate profits as a percentage of GDP (Federal Reserve)*

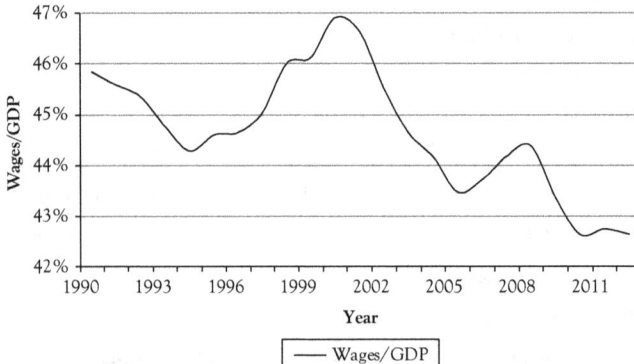

Figure 8.4 *Wages as a percentage of GDP*
Source: Federal Reserve (http://research.stlouisfed.org/fred2).

90s, the share of the top 1 percent grew strongly. The share of income for the top 1 to 5 percent group and the top 5 to 10 percent also grew but at a much slower rate than for the top 1 percent. Following the Great Recession, Saez points out that "the top 1 percent captured 95 percent of the income gains in the first three years of the recovery" so that the "top 1 percent incomes are close to full recovery while bottom 99 percent incomes have hardly started to recover." One clear implication of Figure 8.5 is that for the top 1 percent to grow from 10 percent of U.S. income in 1980 to 22.5 percent in 2012, the share for the lower 99 percent had to fall from 90 percent to 77.5 percent. The top 10 percent group has grown from 36.4 percent of U.S. income in 1980

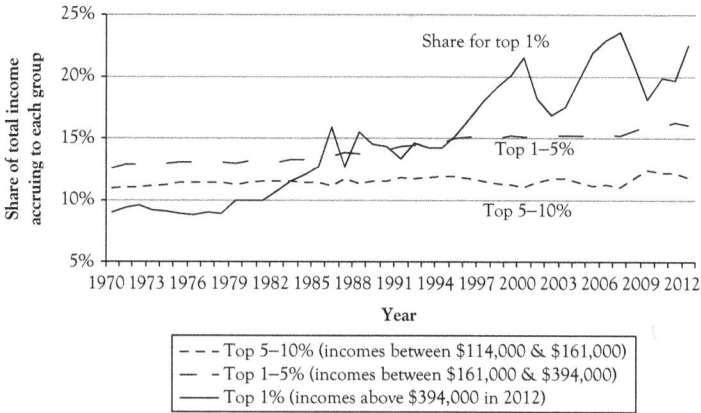

Figure 8.5 U.S. income share for top 1 percent, 1–5 percent, and 5–10 percent

to 50.5 percent in 2012. In other words, the top 10 percent in income now get over half of total U.S. income. From 1993 to 2012, the real income of the top 1 percent grew by 86 percent while the real income of the bottom 99 percent grew by only 6.6 percent (Saez 2013). So-called "Free Trade," as is currently practiced by the United States, seems only to be working for the very wealthy.

2014

It has been 20 years since NAFTA was enacted and 14 years since China was granted PNTR. Over these years, the United States has built up huge trade deficits and millions of jobs have been lost. As presented in this chapter, warnings and alarms have been given every year regarding the damage being done to U.S. manufacturing, to American families, communities, and national security. Yet in 2014, a paper was published with perhaps the most ironic title ever "The Surprisingly Swift Decline of U.S. Manufacturing Employment" (Pierce and Schott 2014). Surprising to whom?—it's been going on for 20 years. The two authors, one from the Federal Reserve and the other from Yale, conducted an exhaustive statistical and modeling effort that, according to them, found "a link between the sharp drop in U.S. manufacturing employment beginning in 2001 and a change in U.S. trade policy that eliminated potential tariff increases on Chinese imports." You think?

It may be surprising to academics and folks within the Beltway, but the American public is neither surprised nor fooled. As attributed to Abraham Lincoln, "You can fool some of the people all of the time, and all of the people some of the time, but you cannot fool all of the people all of the time." The American people are not fooled. A recent poll conducted by the Mellman Group and North Star Opinion Research (Mellman Group and North Star Opinion Research 2014) found "By more than a 2-1 margin, voters consider outsourcing, rather than a potential shortage of skilled workers, as the reason for a lack of new manufacturing jobs (65 percent vs. 28 percent). Moreover, U.S. citizens see that much of the problem lies with China. "By a 2-1 margin (60 percent vs. 30 percent), voters say the U.S. needs to 'get tough' with countries like China in order to halt 'unfair trade practices, including currency manipulation, which will keep undermining our economy.' " Voters also do not have fanciful notions that everything will be rosy and manufacturing will come back on its on—"A full 50 percent of voters also believe more manufacturing jobs are leaving the country than returning, and only 13 percent believe there is a 'reshoring' trend." Job creation is named as the top priority for voters "of every political affiliation." Scott Paul, President of Alliance for American Manufacturing, said of the poll and needed actions, "Far bolder actions than what many inside the Beltway seem to think: strong Buy American provisions, investments in our workforce, getting tougher with China, and enforcing our trade laws." A recent Gallup poll (Riffkin 2014) reinforced these concerns and findings. Nearly one in four Americans thought that jobs and unemployment was the most important issue facing the country. Voters of all three political leanings thought jobs were the top issue: Democrats 24 percent, Republicans 24 percent, and Independents 23 percent.

You can fool some of the people all of the time, and all of the people some of the time, but you cannot fool all of the people all of the time.

What to Do, What to Do?

In theory, there is no difference between theory and practice; in practice there is.

Attributed to Yogi Berra

In theory, free trade between two countries is positive for both. In practice, that theory has not worked out so well for the United States "Free" trade as practiced today is far from the theory of free trade. Non-tariff barriers to entry, government favoritism and subsidies for State-owned companies, subsidized production of exports, currency manipulation, and so forth all limit and constrain the degree of freedom in "free trade." The United States has played by the rules laid out in various trade agreements only to see other countries fail to comply with the terms of the trade pacts. As a result, the United States has huge deficits, has lost millions of jobs, has seen many industries nearly disappear, and has an enormous dependency on China for critical manufactured products. The threat now is a surge of protectionism as perhaps indicated in the polls described earlier in this chapter. Rampant protectionism would be a mistake. The United States must remain committed to free trade and free markets, but it must insist on compliance with trade agreements.

The goals of this book were: to examine the health of U.S. manufacturing on an industry-by-industry basis; to analyze the declines in employment and the number of plants; to identify the likely drivers of the drops in employees and plants; and to discuss possible government policies. Issues that must be addressed in the development of a comprehensive policy strategy must include:

Undervaluation of the Chinese renminbi (RMB)
Overvaluation of the U.S. dollar
Flooding of U.S. markets with artificially low priced and subsidized products
Competitiveness of U.S. manufacturing industries
Loss of strategic U.S. manufacturing industries
Non-tariff barriers to Chinese markets for U.S. manufacturers
Innovative manufacturing technologies
Workforce training

Currency

Exchange rates measure the value of one country's currency relative to the currencies of other countries. Specifically, the exchange rate represents the rate at which one currency can be exchanged for another. For example, on March 18, 2014, one U.S. dollar (USD) could be exchanged for 0.718 euro (EUR), 0.6016 British pounds (GBP), and 6.195 Chinese yuanrenminbi (CNY) (www.xe.com). Exchange rates are used to calculate the value of foreign goods, services, and assets between one currency and another.

Exchange rates generally float in a market system responding to the relative supply and demand for currencies in foreign exchange markets. The relative demand for currencies is determined by the demand for goods, services, and assets denominated in each currency. For example, if there is high demand in the United States for goods made in China and denominated in Chinese yuan, one would expect this demand to drive the value of the yuan upwards relative to the dollar, that is, it would appreciate in value. This would mean fewer CNY per dollar. However the CNY does not float freely. Figure 8.6 presents historical data for CNY to USD (OANDA n.d.). In 1994, the Chinese government set or "pegged" the yuan at 8.7 yuan to the dollar. It was allowed to rise to 8.28 in value relative to the dollar by 1997 and then remained relatively constant until July 2005 (Morrisonand Labonte 2013). This was, of course, a time of soaring demand for Chinese goods and imports in the United States yet the yuan did not appreciate. As Scott notes "The Chinese central bank

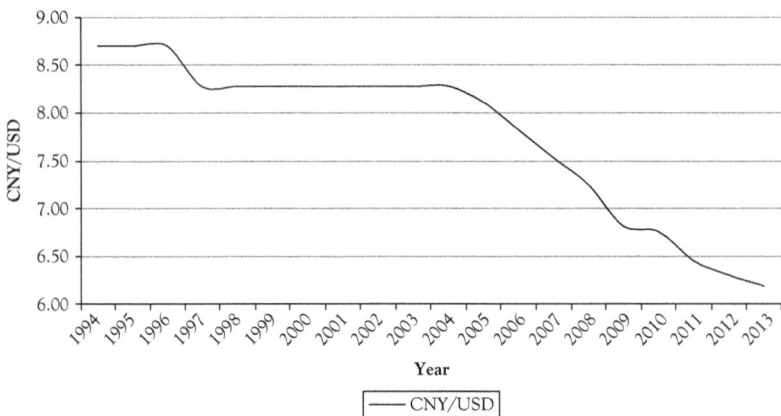

Figure 8.6 Exchange rate for Chinese yuan renminbi (CNY) to U.S. dollar

Source: OANDA (March 12, 2014).

maintained this peg by buying (or selling) as many dollar-denominated assets in exchange for newly printed yuan as needed to eliminate excess demand (supply) for the yuan" (Scott 2014). This kept the price of Chinese imports low and the price of U.S. exports to China high.

Figure 8.6 can, at times, be confusing to interpret because a downward movement represents an appreciation of value for the yuan. Figure 8.7 inverts the vertical axis so that movement up on the figure is an increase in value or in appreciation. In mid-2005, China began to allow the yuan to appreciate but very slowly. By mid-2008, the yuan had appreciated from 8.11 to 6.83. Because of the Great Recession and falling demand for Chinese goods, the Chinese government intervened to halt the appreciation of the yuan. From mid-2008 to mid-2010, the yuan was held more or less constant around 6.83. In 2009, Ferguson and Schularick estimated that the yuan was "clearly undervalued by somewhere between 30 and 48 percent ... and there are good reasons to believe that these estimates are likely to mark to the lower bound."

In mid-2010, as the recovery progressed, the People's Bank of China again let the yuan begin a slow appreciation. By the second week of January 2014, it had appreciated to 6.0904 (average from Monday 1/13/2014 to Sunday 1/19/2014 OANDA n.d.). This represents an appreciation of only 11 percent. However, in mid-January, the yuan began to lose

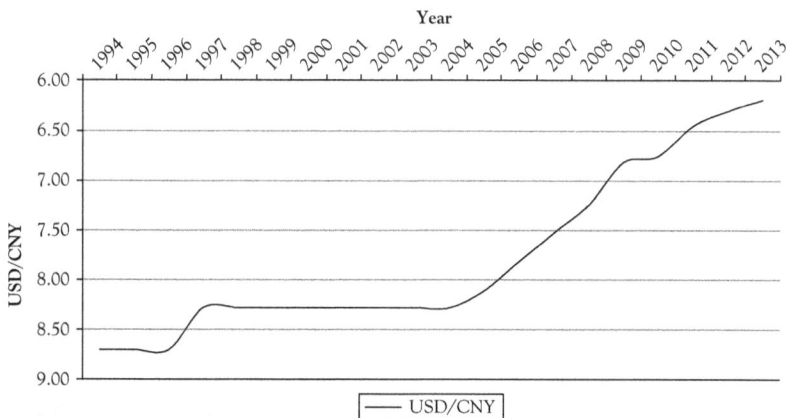

Figure 8.7 Exchange rate for U.S. dollar to Chinese yuan renminbi (CNY)

Source: OANDA (n.d.).

value and depreciate as shown in Figure 8.8. The yuan peaked in value at 6.0402 on January 14, 2014 and had fallen to 6.1258 by March 7, 2014, a depreciation of 1.42 percent (Federal Reserve n.d.). This depreciation again raises concerns of currency manipulation, cheap imports for the United States, and expensive exports to China. As Cline recommends "it is important that China change its peg from the dollar to a basket to stabilize the effective rate" (Cline and Williamson 2009). In a 2012 paper, Cline further observes that "China still has fast productivity growth in the tradable goods industries, which implies that a process of continuing appreciation is essential to maintain its current account balance at a reasonable level" (Cline and Williamson 2012).

In a follow-up study, Cline explored what he termed "Aggressive-Rebalancing" (Cline 2013). In this analysis, Cline defined Aggressive-Rebalancing as:

> *Rich* countries would be expected to have a current account balance of *at least zero*, but with the 3 percent of GDP ceiling used in the standard FEERs calculations. Correspondingly, *developing and emerging market* economies would be expected to have a current account balance of *at most zero*, and again with a lower bound of –3 percent of GDP as in the standard method.

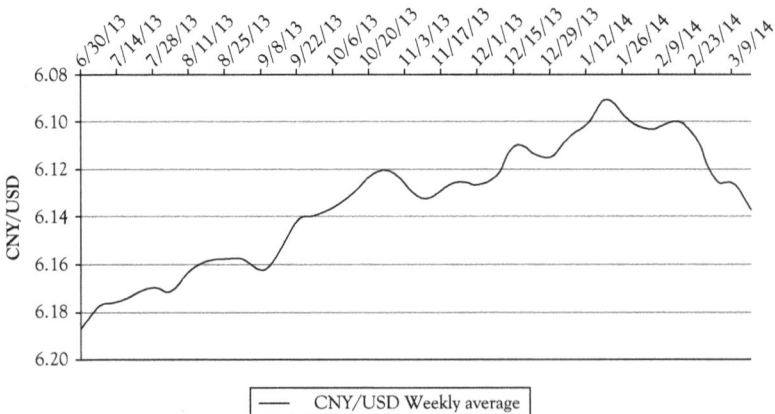

Figure 8.8 Exchange rate for Chinese yuan renminbi (CNY) to U.S. dollar 6/30/13–3/16/14

Source: Federal Reserve (n.d.).

In this scenario, Cline found that "China's currency rises 31 percent bilaterally against the dollar to reach the Aggressive Rebalancing FEER-consistent level." Whatever the required yuan appreciation is, fair and equilibrium exchange rates must be a key policy to save American manufacturing.

Competitiveness Audit

Mandel and Carew have proposed an important activity to help save American manufacturing—a competitiveness audit (Mandel and Carew 2011). Although millions of U.S. jobs have been lost, there is still potential, under fair and policed trade pacts, of growing markets for U.S. goods and services in the developing world. But as Mandel and Carew point out, "In this global economy, we need to know which industries are internationally competitive, which ones aren't, and whether the gaps are closing or widening." Unfortunately, we don't have this data, so developing a well-targeted strategy is not really possible.

Currently the BLS collects much data on *changes* in import and export prices in the International Price Program. For example, from February 2013 to February 2014, U.S. export prices for Transportation Equipment (NAICS 336) increased by 1.4 percent, while over the same period, U.S. import prices declined by 0.8 percent (Bureau of Labor Statistics 2014a, 2014b). The BLS, however, does not directly compare the *level* of import and export prices with the *level* of domestic producer prices. For example, it does not compare the price of imported circuit boards with the price of a domestically produced circuit board with near identical specs nor does it compare the price of a specific domestically produced auto part with its imported counterpoint. Without this type of data, it is impossible to know which industries are globally competitive, which are nearly competitive, and which are very noncompetitive. These competitive positions must be known in order to develop government strategies regarding investments in technology and training as well as trade policies. A competitiveness audit is an essential action needed in the short term to save American manufacturing.

Trade Policy

As a leading member of WTO and a long advocate of free trade, the United States must remain committed to free trade and global cohesiveness. Job losses and industry shrinkage, however, inevitably lead to calls for protectionist measures such as tariffs on certain imported goods. It must be noted that, in the past, unilateral protectionist actions by the United States have generally been failures.

Alexander (Alexander and Soukup 2010) provides two fascinating examples of trade disputes in which a threat of retaliation or retaliation itself led the United States to back off unilateral actions. In 2002, in response to pressures from the U.S. steel industry and to garner support in the important swing states of Pennsylvania, Ohio, and West Virginia, President Bush announced tariffs up to 30 percent on imported steel for three years. The EU, Japan, Korea, China, Switzerland, Norway, New Zealand, and Brazil filed actions with the WTO seeking removal of the tariffs for alleged violation of WTO agreements and articles. A WTO panel found that the U.S. tariffs were inconsistent with WTO rules and requested that the United States remove the tariffs. The United States appealed the findings, but the WTO Appellate Body upheld the results of the panel. The EU threatened to retaliate if the United States did not abide by the WTO findings and remove the tariffs. If a WTO member refuses to implement WTO findings, then the WTO can authorize retaliation by the complaining states in order to induce compliance. In its threatened retaliation, the EU created a carefully targeted list of industries located in important swing states. For example, the list included 100 percent tariffs on fruit juices from Florida, t-shirts from South Carolina, and apples from Washington, again important swing states. These industries and states rapidly began lobbying, and the risk of losing these states led Bush to capitulate and remove the tariffs. Just the threat of retaliation was enough to force compliance with WTO rulings.

The second example provided by Alexander (Alexander and Soukup 2010) regards Mexican trucks in the United States. Prior to NAFTA, Mexican trucks were allowed in the United States only in narrowly defined commercial zones near the border, generally extending only 3 to 20 miles inside the border. Within these zones, the cargo would then need to be transferred to a U.S. truck for delivery into the United States. This of course resulted in delays and higher costs. NAFTA, enacted in 1994, provided a two-step process for easing the trucking regulations. The first step was that Mexican trucks would have complete access to roadways in the border states of California, Arizona, New Mexico, and Texas by December 18, 1995. The second step would enable Mexican trucks to travel freely throughout the United States by January 1, 2000. These provisions were strongly opposed by the Teamsters, environmentalists, and safety and security groups. On December 15, 1995, three days before Step One was to become effective, President Clinton ordered that Mexican trucks would not be allowed beyond the existing commercial zones. Mexico filed a grievance and in 2001, the NAFTA arbitration panel ruled in favor of Mexico and recommended that the United States comply with the provisions of NAFTA. If the United States refused to comply, provisions in the NAFTA articles permit Mexico to impose sanctions against the United States. The NAFTA arbitration panel did say that the United States had the right to set safety regulations for trucks operating within its borders. In 2001, Congress passed legislation with more restrictive language for Mexican trucks, and the legislation was signed into law by President Bush. Mexican trucks that complied with the new requirements could receive temporary permits. Thereafter, a series of legal challenges and the 2004 election delayed implementation of the law. In 2007, President Bush announced a pilot program that would allow 100 Mexican trucks to haul non-hazardous cargo throughout the United States if the drivers were licensed, insured, and could read and speak English. The pilot program was initially for one year but was extended for two additional years. The Teamsters, the Sierra Club and other environmental groups, and Democrats in Congress strongly criticized the Bush administration.

The FY 2009 Omnibus Spending Bill passed by Congress and signed into law by President Obama cut off funding for the pilot program. In response to the ending of the pilot program, Mexico responded strongly and imposed daunting tariffs on 90 products from the United States. The tariffs ranged from 10 to 45 percent. Moreover, Mexico "imposed rotating tariff hikes on a cross section of goods in order to affect the greatest potential impact on U.S. trade and production" (Conkey 2009). The goal was to impose costs on export groups that could bring maximum pressure on political leaders. And it worked as noted in the Washington Post (Booth and Wilson 2009): "In a letter to Obama, 150 U.S. corporations, including General Electric and Wal-Mart, warned: 'The retaliation is already impacting the ability of a broad range of U.S. goods to compete in the Mexican market, from potatoes and sunscreen to paper and dishwashers.' " Tariffs were placed on a wide range of products including onions, pears, cherries, almonds, potatoes, soy sauce, mineral water, strawberries, grapes, wine, shampoo, toothpaste, deodorants, coffee makers, sunglasses, beer, batteries, and Christmas trees (Conkey et al. 2009). Targeted states and politicians included California (grapes, fruit, and wine; House Speaker Nancy Pelosi, Senators Boxer and Feinstein), Ohio (Proctor and Gamble, Smuckers; Senator Sherrod Brown), Oregon (Christmas trees, frozen potatoes, cherries and pears; two Democratic Senators and four Democratic Congressmen), and Connecticut (Duracell batteries; Democratic Senators and Congressmen) (Alexander and Soukup 2010).

After lengthy bilateral negotiations, the Obama administration announced a new pilot program to allow long-haul Mexican trucks into the United States in April 2011. The first Mexican truck heading for the interior of the United States crossed the border on October 21, 2011 (Forsyth 2011). The truck, heading to the Dallas area, was carrying electronic equipment. One hour before the truck left the Nuevo Laredo departure point, Mexico lifted the tariffs it had placed on over 90 U.S. products. Business leaders on both sides of the border were pleased; unions and the Teamsters were not. Twenty years after NAFTA was signed, the dispute and arguments continue.

These two examples of trade disputes are given to highlight several points: (1) unilateral actions by the United States in terms of protectionist tariffs or quotas will surely result in retaliation by the impacted country; (2) the U.S. industries and labor that are being protected will support the protectionist actions; industries that make use of the imported goods will oppose the action; (3) the retaliatory action will target industries in the states of political leadership and those politicians supporting the protectionist stance; and (4) lobbyists of the industries adversely impacted by the retaliation will bring quick and substantial pressure to bear. Ultimately it seems that unilateral protectionist actions are doomed to prolonged disputes and ultimately failure in an elected democracy. The political pressures are simply too strong.

The alternative is aggressive action and the filing of grievances with the WTO. This process, however, is quite lengthy. President Bush filed five grievances against China as shown in Table 8.4: one in 2004, which was resolved in seven months; one in 2006, which was resolved in three years; and three in 2007, one resolved in one year, one resolved in three years, and one resolved in over six years although a MOU has been developed. The compliance on the fifth dispute is still being resolved. All five grievances were decided in favor of the U.S. complaint, and China agreed to implement the necessary changes to be in compliance with WTO rules and procedures (World Trade Organization 2014).

President Obama has filed 10 grievances as shown in Table 8.5. Of the 10 disputes, 6 have not been resolved even though considerable time has passed, ranging from one-and-a-half years to over five years. With U.S. manufacturing jobs being lost every day, these delays are very damaging and the United States must press for faster action by the WTO and China.

Table 8.4 WTO disputes filed under President George W. Bush, the United States as complainant and China as respondent

	Date Filed	Date Resolved
1. Value Added Tax on Integrated Circuits	18-Mar-04	5-Oct-04
2. Measures Affecting Imports of Automobile Parts	30-Mar-04	1-Sep-09
3. Certain Measures Granting Refunds, Reductions or Exemptions from Taxes and Other Payments	2-Feb-07	19-Dec-07
4. Measures Affecting the Protection and Enforcement of Intellectual Property Rights	10-Apr-07	8-Apr-10
5. Measures Affecting Trading Rights and Distribution Services for Certain Publications and Audiovisual Entertainment Produtcs	10-Apr-07	Not Resolved

Table 8.5 WTO disputes filed under President Barack Obama, the United States as complainant and China as respondent

	Date Filed	Date Resolved
Measures Affecting Financial Information Services and Foreign Financial Information Suppliers	3-Mar-08	4-Dec-08
Grants, Loans and Other Incentives	19-Dec-08	Not Resolved
Measures Related to the Exportation of Various Raw Materials	23-Jun-09	1-Jan-13
Certain Measures Affecting Electronic Payment Services	15-Sep-10	19-Aug-13
Countervailing and Anti-Dumping Duties on Grain Oriented Flat-rolled Electrical Steel from the United States	15-Sep-10	Not Resolved
Measures Concerning Wind Power Equipment	22-Dec-10	Not Resolved
Anti-Dumping and Countervailing Duty Measures on Broiler Products from the United States	20-Sep-11	***
Measures Related to the Exportation of Rare Earths, Tungsten and Molybdenum	13-Mar-12	Not Resolved
Anti-Dumping and Countervailing Duties on Certain Automobiles from the United States	5-Jul-12	Not Resolved
Certain Measures Affecting the Automobile and Automobile-Parts Industries	17-Sep-12	Not Resolved

Dates Obtained from WTO website on 1 April 2014 http://www.wto.org/english/tratope/dispue/dispu status e.htm
*** Scheduled to be resolved 9 July 2014

National Security

In addition to employment and economic considerations, there is another vital dimension to manufacturing that must be considered and that is national security. Most computers and peripheral equipment, communications equipment, and audio and video equipment are now imported to the United States from China. This reliance on China for high tech electronics represents not only a huge dependency but a significant vulnerability. In early 2014, the Department of Defense and the General Services Administration jointly released the report "Improving Cybersecurity and

Resilience through Acquisition" (Department of Defense and General Services Administration 2014). The report notes that "Movement of production outside the United States has also led to growing concerns associated with foreign ownership, control, manipulation, or influence over item that are purchased by the government and used in or connected to critical infrastructure or mission essential systems." The report, somewhat ominously continued, "The Federal government and its contractors, subcontractors, and suppliers at all tiers of the supply chain are under constant attack, targeted by increasingly sophisticated and well-funded adversaries seeking to steal, compromise, alter or destroy sensitive information. In some cases, advanced threat vectors target businesses deep in the government's supply chain to gain a foothold and then 'swim upstream' to gain access to sensitive information and intellectual property" (Department of Defense and General Services Administration 2014). There are many reasons for the United States to recapture manufacturing of computers and electronic products. It will require innovation and advanced manufacturing techniques such as printing of electronics, similar to additive or 3D manufacturing. But innovation is what Americans do and must do more of in the future.

Competitiveness and Technology

In order to be globally competitive, American manufacturing must be competitive in cost, quality, and time to market. Of these three, cost and time to market are the critical objectives for American manufacturing. An analysis conducted by the Boston Consulting Group (BCG) indicates that China's cost advantage is eroding for several reasons (Sirkin et al. 2011):

> Wage and benefit increases of 15 to 20 percent per year at the typical Chinese factory will reduce China's labor-cost advantage over low cost states in the United States from 55 percent today to 39 percent in 2015, when adjusted for higher productivity of U.S. workers. Since labor costs are typically a small percentage of product costs, outsourcing to China will provide only single digit savings for many products.

When total landed costs are calculated including transportation, inventory, supply chain risks, duties, and so forth, the cost savings of manufacturing in China as opposed to low cost states in the United States will become minimal; although the renminbi has recently weakened, the general strengthening trend reinforces the loss of cost savings.

Although productivity is growing in China through automation and other means, it will not be enough to maintain a significant cost advantage.

Growing domestic demand in China and Asia will consume more and more of China's production, supporting the move of some production back to the United States for North American consumption.

The potential of moving some China production to other low cost countries such as Vietnam, Indonesia, and Myanmar will be constrained due to concerns about safety, infrastructure, corruption, and low levels of worker skills.

BCG summarizes the implications of these trends as follows: "When all costs are taken into account, certain U.S. states, such as South Carolina, Alabama, and Tennessee, will turn out to be among the least expensive production sites in the industrial world. As a result, we expect to see companies to begin building more capacity in the U.S. to supply North America."

Moser, founder of the Reshoring Initiative, provides a concise list of why companies should reshore to the United States (Moser 2014):

Reduces total cost of ownership;
Improves quality and consistency of inputs;
Reduces pipeline and surge inventory impact on just-in-time operations;
Clusters manufacturing near R&D facilities, enhancing innovation;
Reduces intellectual property and regulatory compliance risk;
Eliminates the waste and instability caused by offshoring; and
Strengthens companies' ability to respond quickly to customers' demands.

To aid companies in analyzing the economics of reshoring manufacturing to the United States, the Reshoring Initiative provides an on-
line calculator "The Total Cost of Ownership Estimator™" that enables
companies to estimate the total cost involved in off-shoring their particular product. The Estimator incorporates 29 cost factors, automatically calculates freight rates for 17 countries, automatically calculates
the duty rate for parts or tools, and provides the current total cost of
ownership value, as well as a five-year forecast based on the user's forecast of wage and currency changes. It is a valuable tool for calculating
the complex total costs of offshoring versus onshoring. The calculator is
available at (Reshoring Initiative 2014).

Technology will play a key role in rebuilding American manufacturing. President Obama's proposed National Network for Manufacturing Innovation (NNMI) is an important part of America's technology
strategy (Advanced Manufacturing National Program Office 2014).
NNMI will consist of regional hubs that will accelerate development
and adoption of cutting-edge manufacturing technologies for making
new, globally competitive products. To date, four institutes have been
announced: National Additive Manufacturing Institute, Digital Manufacturing and Design Innovation Institute, Lightweight and Modern
Metals Manufacturing Institute, and Next Generation Power Electronics Manufacturing Innovation Institute. For maximum impact, these
institutes must focus on reducing costs and product development time,
and increasing innovation in those industries targeted by the competitiveness audit.

Energy

The changing landscape of U.S. energy production and consumption is
highly favorable to rebuilding U.S. manufacturing. A Citi GPS report
(Morse 2012) states that North America has become the fastest growing oil and gas producing region in the world, and is likely to remain
so for the rest of the decade and into the 2020s. The Citi report and a
report from the University of California in Davis (Jaffe 2013) suggest that
price of a barrel of oil could fall from the current level of $100 to $75.
Additionally, the price of natural gas has fallen substantially and is now

roughly one-third the cost of oil in BTU equivalents. Citi sees the abundant natural gas triggering an industrial revolution in energy-intensive industries, as well as shifts to gas-fired power generation, natural gas vehicles, and LNG exports. Each of these developments creates substantial opportunity for American manufacturing.

Final Observations

There now exists in the United States a considerable opportunity and a foundation for growth:

Nearly 12 million employees still work in manufacturing in the United States;

- There are over 340,000 manufacturing establishments in the United States;
- The United States still accounts for 19.4 percent of global manufacturing value added, a share that has declined only slightly over the past three decades;
- The foundation is in place for growth, economics for off-shoring are reversing, and with appropriate government policies, the growth in the manufacturing sector and employment could be sizeable.

U.S. manufacturers, however, are now competing against China Inc., Taiwan Inc., Singapore, and South Korea Inc.—nations as enterprises. The playing field is not level; selected U.S. government policies and programs can play a significant role in enhancing our global competitiveness. Specifically programs aimed at improving productivity and innovation at small to medium size firms will help create a robust supplier network to support new plants. For larger firms considering reshoring, initiatives such as State and Federal programs for plant construction; rapid depreciation of plant and equipment; government support for applied R&D; lower tax rates on manufacturing facilities; and training of new employees will support the economics of the reshoring decision and building anew in the United States. The opportunities are real. We must, and we can, save American manufacturing.

With millions of jobs lost and thousands of factories closed, it should be clear that, as a nation, we are losing the global manufacturing war. It is not, however, the skill and commitment of American workers, the quality of American products, or the lack of innovation that limits our ability to compete successfully. The current manufacturing and economic crisis has been created by our political process that is far too easily manipulated by money. Large U.S. multinationals and foreign countries, their lobbyists, and their campaign contributions have a dominant influence over a Congress focused more on re-election and securing campaign funds than jobs and national security. With the loss of manufacturing, there has been a dramatic decline in union membership and a related loss of pro-jobs lobbying. The results are trade agreements that are disastrous for both American workers and national security but highly profitable for the corporations and economically beneficial to the foreign countries. As noted in a New York Times article in 2004, "American jobs are being lost not to competition from foreign companies, but to multinational corporations, often with American roots, that are cutting costs by shifting operations to low-wage countries" (Schumer 2004). David E. Bonior was the House Democratic whip at the time of the NAFTA vote. He recently wrote in a New York Times Op-Ed column, "The companies that took the most advantage of NAFTA—big manufacturers like G.E., Caterpillar and Chrysler—promised they would create more jobs at their American factories if NAFTA passed. Instead, they fired American workers and shifted production to Mexico" (Bonior 2014). The drives for increased profits and reduced environmental, worker health and safety regulations motivate intense political activities by the multinationals and the foreign countries. Nobel Laureate Joseph E. Stiglitz puts it like this, "Corporations everywhere may well agree that getting rid of regulations would be good for corporate profits. Trade negotiators might be persuaded that these trade agreements would be good for trade and corporate profits. But there would be some big losers—namely, the rest of us" (Stiglitz 2014).

A good example illustrating lobbying and political pressure on Congress is the recently enacted Korea, Panama and Columbia free trade agreements. Of the three countries, the agreement with Korea drew the most attention due to several areas of concern.

First was the concern that the United States might end up importing goods made at the Kaesong Industrial (KIC) Park in North Korea.
The KIC is a six-year-old industrial park located in the Democratic People's Republic of Korea (DPRK or North Korea) just across the demili-
tarized zone from South Korea. The complex was planned, developed,
and financed largely by South Korea, and it has become a symbol of
engagement between the North and the South. KIC is operated by
the South Korean company Hyundai Asan. According to the Congressional Research Service (Manyin 2011), as of the end of 2010, over 120
medium-sized South Korean companies were employing over 47,000
North Korean workers to manufacture products in Kaesong. The products vary widely, and include clothing and textiles (71 firms), kitchen
utensils (4 firms), auto parts (4 firms), semiconductor parts (2 firms), and
toner cartridges (1 firm). Conceivably these products could make their
way into South Korean products and be imported to the United States in
violation of the U.S. import ban on products from North Korea. A second troubling aspect of the Korean free trade agreement is that Korean
products containing up to 65 percent Chinese content can be shipped
duty-free into the United States, in effect, legalizing trans-shipments that
otherwise would face duties. A third concern is that EPI forecasts that
the agreement will result in the loss of 159,000 jobs in the United States
(Scott 2010). Fourth and finally, the KIC is a major source of hard currency for the North Korean regime, supporting military activities and
the lavish lifestyle of Kim Jong-un. Wages of the North Korean workers, which average about $60 a month, are paid directly to the North
Korean government which withholds around 45 percent (Manyin 2011).
As noted by the New York Times, the North also makes money by selling
50-year leases on the land in KIC (Fackler 2008).

Because of these four concerns and troubling issues, lobbying activity went into high gear to overcome any potential opposition. A review
by *The Hill* of Justice Department records shows that the Washington
embassies, foreign ministries and trade agencies of Korea, Columbia
and Panama spent at least $15 million on lobbying, legal and PR work
between the beginning of 2006 and October of 2011 to press for passage
of the free-trade agreements (FTA) (Trade Deals Were Cash Cow for K
Street 2011). The Korean government was the largest spender among the

three with $6.3 million spent on lobbying and PR. Big business lobbying was led by the U.S.-Korea FTA Business Coalition whose president was also the vice-president, Asia for the U.S. Chamber of Commerce (Froomkin 2011). Members include a who's-who of U.S. multinationals including General Electric, Caterpillar, GM Korea, Pfizer, Merck, Pharmaceutical Research & Manufacturers of America, and Citigroup (U.S.-Korea Business Council 2014).

The lobbying was intense and effective. Even with all the significant concerns, when the Senate Finance Committee took up the free trade agreements, not a single member of the committee, Republican or Democrat, made a negative comment about the deal (R. McCormack 2011). The lobbying and campaign contributions also contained an implied threat regarding the money: "On the one hand, you do what I want and I will keep giving it to you. On the other hand, if you don't, I will use copious amounts to make your next election a misery" (Froomkin 2011). Congress passed the Korea, Columbia and Panama free trade agreements on October 12, 2011 by a rather lopsided vote.

Since the passage of NAFTA 20 years ago, the United States has been caught in a vicious cycle spiraling downward. With each passing year, U.S. manufacturing has grown weaker and our dependency on foreign imports has grown. The charts in Chapter 7 presented the on-going and rapid growth in trade deficits for a wide array of manufactured goods. These charts show the real world results of the vicious cycle. Nobel Laureate Joseph E. Stiglitz terms this "our gross mismanagement of globalization." Today, the United States is especially dependent upon China for electronics, computers and peripheral equipment, and communications equipment. With our growing dependency, our ability to protest and correct non-compliance to free trade agreements is greatly diminished. We are almost too weak to put up a fight—the damaging impacts of retaliation are great. Yet with each passing year the United States becomes more dependent and less able to demand true "free trade." The United States must react before it is too late, and it must take a strong and forceful stance.

The danger is the nature of that stance. The American public is increasingly hostile to free trade. A 2010 poll conducted by the Wall Street Journal and NBC News (Murray 2010) found "83 percent of blue-collar workers agreed that outsourcing of manufacturing to foreign countries with lower wages was a reason the U.S. economy was struggling and more people weren't being hired; no other factor was so often cited for current economic ills. Among professionals and managers, the sentiment was even stronger: 95 percent of them blamed outsourcing." 84 percent of Democrats agreed and 90 percent of Republicans agreed. The danger here is that of a populist appeal to unilateral protectionist measures that could ignite a trade war. That is not a promising approach for the United States.

Recommended Actions

The United States must take forceful action in five areas:

1. The United States must insist that other countries stop manipulating their currencies and permit the dollar to regain a competitive level. The United States should first seek voluntary agreement from the currency manipulators to greatly reduce or eliminate their intervention. If the manipulators do not do so, however, the United States should adopt four new policy measures as suggested by Bergsten and Gagnon (Bergsten 2012):

    (1) Undertake countervailing currency intervention (CCI) against countries with convertible currencies by buying amounts of their currencies equal to the amounts of dollars they are buying themselves, to neutralize the impact on exchange rates,

    (2) Tax the earnings on, or restrict further purchases of, dollar assets acquired by intervening countries with inconvertible currencies (where CCI could therefore not be fully effective) to penalize them for building up these positions,

    (3) Treat manipulated exchange rates as export subsidies for purposes of levying countervailing import duties, and

    (4) Hopefully with other adversely affected countries, bring a case against the manipulators in the World Trade Organization that would authorize more wide-ranging trade retaliation.

        Bergsten and Gagnon suggest this approach should first be taken against eight of the most significant currency manipulators: China, Denmark, Hong Kong, Korea, Malaysia, Singapore, Switzerland, and Taiwan.

2. The United States must pursue aggressive actions and complaints with the WTO. In 2013, in a very difficult trade environment, the United States filed not a single complaint with the WTO. Given the extent to which the United States is being "beat up" in international trade, this lack of action is difficult to understand. Moreover, multiple complaints from previous years have not been resolved. "While Nero fiddled ...." The United States must seek resolution of past complaints and continue aggressive filing with this world body.

3. The national government should take a lesson from State governments and actively compete globally for new manufacturing facilities. This competitive stance should include lower income tax rates for manufacturing, tax credits for manufacturing job creation, more rapid depreciation on plant and equipment, permanent R&D tax credits and other incentives. These actions will help level the global playing field.

4. The U.S. Congress should fund the proposed National Network for Manufacturing Innovation. This network will bridge the so-called "valley of death" between basic research and products made in American factories. Technology, innovation and creativity have been the bedrock of U.S. economic growth. These institutes have the very real potential for initiating a renaissance in manufacturing and job creation.

5. The Department of Commerce should immediately undertake a competitiveness audit to identify those industries with America is still globally competitive and near-competitive. This audit will provide valuable insights for investment and strategy decisions. Otherwise, actions will be taken in the dark.

If these five actions are forcefully undertaken, and if Congress can stand up to lobbyists, American manufacturing can be saved, good jobs will be created, income inequality will be reduced, and national security will be strengthened.

Epilogue

I asked my friend Dick Reeves, Managing Partner of Angel Syndicates Central, to review a late draft of this book. Dick is very involved in the financing of start-up companies and the creation of jobs. His response was a thoughtful question to me: "Bill, are things really as bad as your data indicate? What about the new jobs being created in apps, social media, etc.? The unemployment rate is falling, so aren't things basically alright?"

That question sent me back into analysis mode. As Dick indicated, the unemployment rate has indeed been falling. In May of 2014, the unemployment rate had fallen to 6.3 percent from a high of 10 percent in October, 2009 during the height of the recession (Bureau of Labor Statistics 2014). Chart 1 illustrates the slow, but steady, drop in the unemployment rate toward the average rate of 5.44 percent that existed between 1990 and 2007. Does this drop really mean that, in reality, the loss of manufacturing jobs is "much ado about nothing"? I'm afraid not. A falling unemployment rate only tells one side of the story.

The official unemployment rate includes those unemployed workers who are able to work, have actively looked for work in the past four

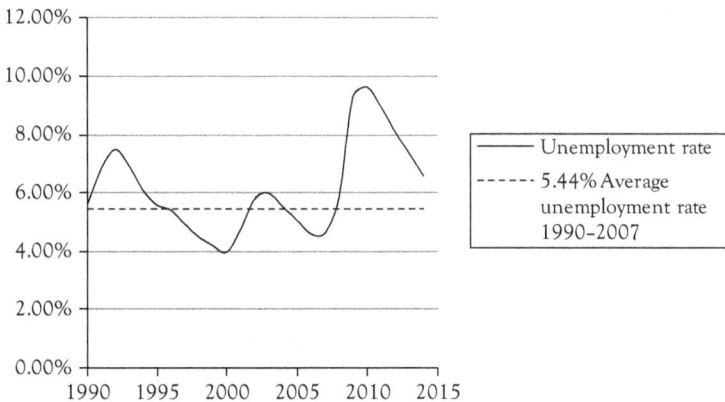

Chart 1 *U.S. Unemployment rate*

weeks, but have not found or taken a job or been recalled to a previous job. A person is not counted as unemployed if he or she has gotten frustrated with looking for a job and has given up trying to find work. This implies that the official unemployment rate understates the true rate of unemployment. Moreover, the unemployment rate can also understate the actual unemployment rate because it does not include those who are underemployed, for example, working part-time when they would prefer to be working full-time; or who are working at jobs that are below their skill, educational, or wage levels.

Because the unemployment rate does not include many individuals who have given up or are not fully employed, the labor force participation rate is often considered a more indicative measure of employment. This measure is determined as the percentage of the adult population that is working. More precisely, it is the proportion of the civilian noninstitutional (not in college or the Army for example) population aged 16 years and over that is employed. As may be seen in Chart 2, in the year 2000 roughly 64.5 percent of the population in the United States was working (Bureau of Labor Statistics 2014). That percentage has now fallen to approximately 58.5 percent, a drop of 6 percent. Table 1 provides striking data on this phenomenon (Bureau of Labor Statistics 2014). Between the years 1990 and 2000, the number in the civilian noninstitutional population increased by 12.4 percent. The number employed grew even faster, with an increase of

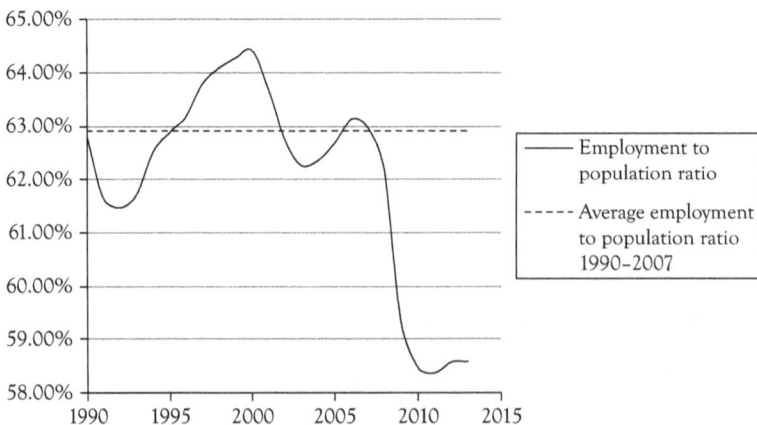

Chart 2 Employment to population ratio

Table 1 Employment status of the civilian noninstitutional population 16 years and over

Year	Civilian Noninstitutional Population	Employed	Not in Labor Force
1990	189,164	118,793	63,324
1991	190,925	117,718	64,578
1992	192,805	118,492	64,700
1993	194,838	120,259	65,638
1994	196,814	123,060	65,758
1995	198,584	124,900	66,280
1996	200,591	126,708	66,647
1997	203,133	129,558	66,837
1998	205,220	131,463	67,547
1999	207,753	133,488	68,385
2000	212,577	136,891	69,994
2001	215,092	136,933	71,359
2002	217,570	136,485	72,707
2003	221,168	137,736	74,658
2004	223,357	139,252	75,956
2005	226,082	141,730	76,762
2006	228,815	144,427	77,387
2007	231,867	146,047	78,743
2008	233,788	145,362	79,501
2009	235,801	139,877	81,659
2010	237,830	139,064	83,941
2011	239,618	139,869	86,001
2012	243,284	142,469	88,310
2013	245,679	143,929	90,290
2014	247,264	145,543	91,575
1990-2000	23,413	18,098	6,670
% Change	12.38%	15.23%	10.53%
2000-2014	34,687	8,652	21,581
% Change	16.32%	6.32%	30.83%

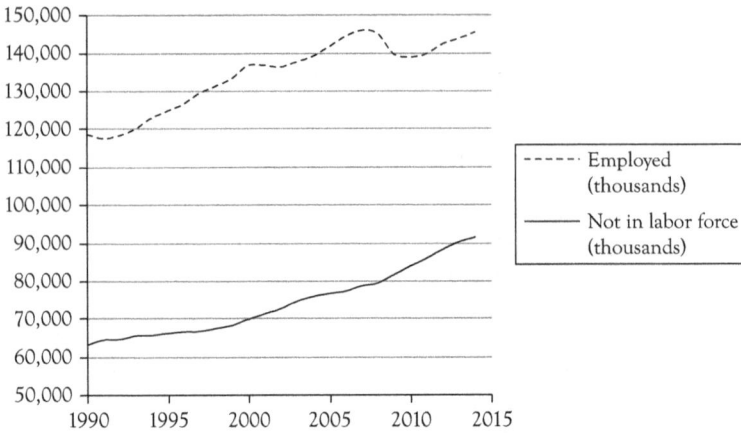

Chart 3 Number employed and number not in the labor force

15.2 percent. The number of individuals not in the labor force grew by just 10.3 percent, lagging behind the growth of the population and the number of employed. In other words, jobs were being created faster than the growth of the civilian noninstitutional population. After the year 2000, however, that dynamic changed rather dramatically. As seen in Table 1, the civilian noninstitutional population grew by 16.3 percent from 2000 to 2014, close to the same rate as from 1990 to 2000. The number employed, however, only grew by 6.3 percent. Over the same period, the number of individuals not in the labor force, the ones being left behind, grew by 21.3 million workers, an increase of 30.8 percent—a huge increase. As seen in Chart 3, the population not in the labor force grew sharply after the year 2000 and now totals 91.5 million citizens of the United States.

The loss of six million manufacturing jobs surely matters. That importance is heightened by the fact that every manufacturing job is estimated to generate from 2.2 to 4.6 additional jobs in support industries (Wial 2013). Even at the low end, that's an additional 12 million jobs lost for a total loss of 18 million jobs, a significant part of the 21 million added to those not in the labor force between 2000 and 2014. It is vital that we save American manufacturing and stop the bleeding of jobs.

About the Author

Dr. William R. Killingsworth is Vice President of Manufacturing and Supply Chain Research at DESE Research, Inc., a high technology consulting and research firm. For 10 years, Dr. Killingsworth was the Executive Director of the MIT Forum for Supply Chain Innovation at the Massachusetts Institute of Technology. He also served as Director of the Office for Enterprise Innovation and Sustainment at the University of Alabama in Huntsville.

Dr. Killingsworth conducts research and consulting in advanced manufacturing, supply chain design and optimization, risk management, product lifecycle management, enterprise software, analytics, and master data management. Clients and research sponsors include government and defense agencies and corporations in the aviation and aerospace, automotive, electric power, and pharmaceutical industries. Over the past 10 years, he has worked extensively in the areas of manufacturing trends, industrial base assessment, and supply chain issues with the Office of the Secretary of Defense, and the DoD Office for Manufacturing and Industrial Base Policy, the Army Aviation and Missile Command, and the NASA National Center for Advanced Manufacturing. Dr. Killingsworth participated in the Sector by Sector, Tier by Tier (S2T2) assessment of the defense manufacturing industry that was conducted by OSD. He has also supported the Defense Wide Manufacturing Technology Office within OSD on a number of projects, with particular emphasis on advanced manufacturing. He has also consulted with aerospace and defense prime contractors and first, second, and third tier aerospace suppliers. Dr. Killingsworth has conducted research in demand planning and forecasting, inventory strategies, configuration management, and integrated production planning. He has been invited to give keynote addresses on manufacturing and supply chain management in Australia, Austria, Germany, Greece, Korea, United Kingdom, and Russia and has spoken at many conferences including the Defense Manufacturing Conference, the Aviation Week Supply Chain Conference, the IDGA Military Logistics Conference, the

AIA Product Support Conference, and the Microsoft High Tech Summit. He is the author of the book *Design, Analysis and Optimization of Supply Chains—A System Dynamics Approach.*

With 30 years of experience in management, supply chain, and manufacturing consulting, he founded Killingsworth Associates—a management consulting firm based in Cambridge, Massachusetts. His firm conducted consulting assignments across the United States, Europe, and Africa, specializing in strategic planning, manufacturing and supply chain design, continuous improvement processes, business case analyses, budgeting and capitalization. For a number of years, Dr. Killingsworth also worked as a Registered Investment Representative for Porter White & Company, an investment banking firm, with capitalization of growing technology companies as his area of focus.

With a BS in Aerospace Engineering, Summa Cum Laude, from Auburn University, Dr. Killingsworth attended MIT for graduate school and was a National Science Foundation Fellow. At MIT, he received the SM in Aeronautics and Astronautics specializing in optimization and control theory; and his PhD was in a joint program of the School of Engineering and the Sloan School of Management. His dissertation focused on the optimization, modeling and simulation of business processes in areas such as supply chain management and operations.

References

Advanced Manufacturing National Program Office. 2014. "National Network for Manufacturing Innovation (NNMI)." http://manufacturing.gov/nnmi.html

Alexander, K.W., and B.J. Soukup. 2010. "Obama's First Trade War: The U.S.-Mexico Cross-Border Trucking Dispute and the Implications of Strategic Cross-Sector Retaliation on U.S. Compliance Under NAFTA." *Berkeley Law Scholorship*, http://scholarship.law.berkeley.edu/cgi/viewcontent.cgi?article =1383&context=bjil

Alliance for American Manufacturing. January 2012. "The Attack on the American Auto Parts Industry." *American Manufacturing*, http://americanmanufacturing. org/files/Auto%20Parts%20White%20Paper%20Final.pdf

American Society of Health-System Pharmacists. 2014. *National Drug shortages*, http://www.ashp.org/DocLibrary/Policy/DrugShortages/Drug-Shortages-Statistics.pdf

Atkinson, R.D., L.A. Stewart, S.M. Andes, and S.J. Ezell. March 2012. "Worse than the Great Depression: What Experts Are Missing About American Manufacturing Decline." *The Information and Innovation Technology Foundation (ITIF)*, http://www2.itif.org/2012-american-manufacturing-decline.pdf

Autor H. David, D. Dorn, and G.H. Hanson. October 2013. "The China Syndrome: Local Labor Market Effects of Import Competition in the United States." *American Economic Review* 103, no. 6, pp. 2121–2168.

Bevins, J. April 8, 2004. "Shifting Blame for Manufacturing Job Loss—Rising Trade Deficit Shouldn't be Ignored." *Economic Policy Institute*, http://www. epi.org/publication/briefingpapers_bp149/

Bevins, J. March 22, 2013. "Using Standard Models to Benchmark the Costs of Globalization for American Workers Without a College Degree." *Economic Policy Institute*, http://s3.epi.org/files/2013/standard-models-benchmark-costs-globalization.pdf

Bloomberg News. November 13, 2013. *U.S. Crude Production Beat Imports in October, EIA Says*, http://www.bloomberg.com/news/2013-11-13/u-s-crude-production-beat-imports-in-october-eia-says.html

Booth, W., and S. Wilson. April 15, 2009. "Obama Prepares for Mexico Talks on Drug Trade." *The Washington Post*, http://www.washingtonpost.com/wp-dyn/ content/article/2009/04/14/AR2009041403224.html

Bronfenbrenner, K. September 30, 1996. "Final Report: The Effects of Plant Closing or Threat of Plant Closing on the Right of Workers to Organize."

Digital Commons ILR School, Cornell University, http://digitalcommons.ilr.
cornell.edu/cgi/viewcontent.cgi?article=1000&context=intl

Bronfenbrenner, K. March 1, 1997. "We'll Close! Plant Closings, Plant-Closing
Threats, Union Organizing and NAFTA." *Digital Commons, ILR School,
Cornell University,* http://digitalcommons.ilr.cornell.edu/cgi/viewcontent.
cgi?article=1018&context=cbpubs

Bronfenbrenner, K., L. Stephanie, R. Hickey, T. Juravich, and J. Burke. June
30, 2001. "Impact of U.S.-China Trade Relations on Workers, Wages,
and Employment: Pilot Study Report." *Digital Commons, ILR School,
Cornell University.* http://digitalcommons.ilr.cornell.edu/cgi/viewcontent.
cgi?article=1038&context=reports

Bureau of Economic Analysis. 2014. *U.S. Economic Accounts,* http://www.bea.gov/;
http://www.bea.gov/iTable/itable.cfm?reqid=51&step=1#reqid=51&isuri=1

Bureau of Labor Statistics. 2013. "Quarterly Census of Employment and Wages."
http://www.bls.gov/cew/dataguide.htm

Bureau of Labor Statistics. 2014a. "Quarterly Census of Employment and
Wages." http://www.bls.gov/cew/cewfaq.htm

Bureau of Labor Statistics. 2014b. "Table 3 U.S. Import Price Indexes, by NAICS"
Economic News Release, http://www.bls.gov/news.release/ximpim.t03.htm

Bureau of Labor Statistics. 2014c. "Table 4 U.S. Export Price Indexes, by NAICS"
Economic News Release, http://www.bls.gov/news.release/ximpim.t04.htm

Centers for Disease and Prevention. 2013. Trends in Current Cigarette Smoking
Among High School Students and Adults, United States, 1965–2011."
Smoking & Tobacco Use, http://www.cdc.gov/tobacco/data_statistics/tables/
trends/cig_smoking/

Chicago Tribune. November 18, 1993. "Labor Lost But Showed Muscle." *Chicago
Tribune,* http://articles.chicagotribune.com/1993-11-18/news/9311180034
_1_nafta-battle-nafta-campaign-labor

Clifford, S. September 19, 2013. "U.S. Textile Plants Return, With Floors Largely
Empty of People." *New York Times,* http://www.nytimes.com/2013/09/20/
business/us-textile-factories-return.html?pagewanted=all&_r=0

Cline, W.R. May 2013. "Estimates of Fundamental Equilibrium Exchange Rates,
May 2013." *Peterson Institute for International Economics*, http://www.iie.
com/publications/pb/pb13-15.pdf

Cline, W.R., and J. Williamson. June 2009. "2009 Estimates of Fundamental
Equilibrium Exchange Rates." *Peterson Institute for International Economics,*
http://www.piie.comwww.piie.com/publications/pb/pb09-10.pdf

Cline, W.R., and J. Williamson. May 2012. "Estimates of Fundamental
Equilibrium Exchange Rates May 2012." *Peterson Institute for International
Economics,* http://www.iie.com/publications/pb/pb12-14.pdf

Clinton, W.J. September 19, 2000. "Clinton Statement on Senate China PNTR
Vote." *FAS (Feberation of American Scientists),* https://www.fas.org/news/
china/2000/prc-000919c.htm

CNN. April 4, 2008. "FDA Thinks it Has Trigger in Heparin Deaths." *CNN Health,* http://www.cnn.com/2008/HEALTH/04/21/fda.heparin/

Commission on U.S.-Pacific Trade and Investment Policy. April 1997. "Building American Prosperity in the 21st Century." *Hathi Trust: Babel,* http://babel.hathitrust.org/cgi/pt?id=mdp.39015041331136;view=1up;seq=3

Congressional Budget Office. February 18, 2004. "What Accounts for the Decline in Manufacturing Employment." *Congressional Budget Office,* http://www.cbo.gov/sites/default/files/cbofiles/ftpdocs/50xx/doc5078/02-18-manufacturingemployment.pdf

Conkey, C., J.De Cordoba, and J. Carlton. March 19, 2009. "Mexico Issues Tariff List in U.S. Trucking Dispute." *The Wall Street Journal,* http://online.wsj.com/news/articles/SB123739445919172781

Dale, R. November 9, 1993. "Thinking Ahead: The U.S. Has Real Reasons to Pass NAFTA." http://www.nytimes.com/1993/11/09/business/worldbusiness/09iht-think_1.html

Department of Defense and General Services Administration. January 29, 2014. "Improving Cyber Security and Resilience Through Acquisition." *Defense,* http://www.defense.gov/news/Improving-Cybersecurity-and-Resilience-Through-Acquisition.pdf.

Dinkelspiel, F. March 27, 2010. "Closing of Auto Plant Forces Suppliers to Scramble." *The New York Times,* http://www.nytimes.com/2010/03/28/us/28sfnummi.html?pagewanted=all&_r=0

Energy Information Administration. 2013. U.S. Refinery and Blender Net production of crude oil and petroleum products." *Petroleum & Other Liquids,* http://tonto.eia.gov/dnav/pet/hist/LeafHandler.ashx?n=PET&s=MTTRPUS1&f=A.

The Economist. May 14, 2011. "Moving Back to America."

Ezell, S.J., and R.D. Atkinson. April 2011. "The Case for a National Manufacturing Strategy." *ITIF (The Information Technology and Innovation Foundation),* http://www2.itif.org/2011-national-manufacturing-strategy.pdf

Federal Reserve. 2014. *Foreign Exchange rates –H.10,* http://www.federalreserve.gov/releases/h10/hist/dat00_ch.htm

The Federal Reserve. November 15, 2013. *G.17, Industrial Production & Capacity Utilization Data,* http://www/federalreserve.gov/datadownload

Ferguson, N., and M. Schularick. December 27, 2007. "Chimerica and the Global Asset Market Boom." *International Finance,* http://onlinelibrary.wiley.com/doi/10.1111/j.1468-2362.2007.00210.x/pdf

Ferguson, N., and M. Schularick. October 2009. "The End of Chimerica." *Harvard Business School,* http://www.hbs.edu/faculty/Publication%20Files/10-037.pdf

Forsyth, J. October 21, 2011. "Years After NAFTA, First Long-Haul Mexican Truck Enters U.S." *Reuters,* http://www.reuters.com/article/2011/10/21/us-trucking-nafta-idUSTRE79K75P20111021

Freeman, R. August 2006. "The Great Doubling: The Challenge of the New Global Labor Market." http://emlab.berkeley.edu/users/webfac/eichengreen/e183_sp07/great_doub.pdf

Fuglie, K., J. MacDonald, and E. Ball. September 2007. "Productivity Growth in U.S. Agriculture, Economic Brief Number 9." *USDA's Economic Research Service*, www.ers.usda.gov; http://www.ers.usda.gov/publications/eb-economic-brief/eb9.aspx#.UxTho-NdXmc

General Accountability Office. September 2008. "Drug Safety: Better Data Management and More Inspections are Needed to Strengthen FDA's Foreign Drug Inspection Program." *GAO*, http://www.gao.gov/new.items/d08970.pdf

General Accountability Office. September 14, 2011. "Drug Safety: FDA Faces Challenges Overseeing the Foreign Drug Manufacturing Supply Chain." *GAO*, http://www.gao.gov/new.items/d11936t.pdf

Goodman, P.S., and J. Healy. March 7, 2009. "Job Losses Hint at Vast Remaking of Economy." *New York Times,* http://www.nytimes.com/2009/03/07/business/economy/07jobs.html?pagewanted=all

Grobart, S. January 5, 2011. "A Bonanza in TV Sales Fades Away." *New York Times,* http://www.nytimes.com/2011/01/06/technology/06sets.html?_r=1&

Harris, G. February 14, 2014. Medicines Made in India Set Off Safety Concerns." *New York Times*, http://www.nytimes.com/2014/02/15/world/asia/medicines-made-in-india-set-off-safety-worries.html

Helper, S., T. Krueger, and H. Wial. February 2012. Why Does Manufacturing Matter? Which Manufacturing Matters? A Policy Framework." *Metropolitan PolicyProgram*, http://www.brookings.edu/~/media/research/files/papers/2012/2/22%20manufacturing%20helper%20krueger%20wial/0222_manufacturing_helper_krueger_wial.pdf

International Center for Trade and Sustainable Development. May 24, 2000. "US House Approves China PNTR." *ICTSD*, http://ictsd.org/i/news/bridgesweekly/88873/

ITA Global Patterns of US Trade. *Global Patterns of U.S. Merchandise Trade*, http://tse.export.gov/TSE/TSEOptions.aspx?ReportID=1&Referrer=TSEReports.aspx&DataSource=NTD (accessed March 1, 2014).

ITA Trade with Selected Market. *Product Profiles of U.S. Merchandise Trade With a Selected Market*, http://tse.export.gov/TSE/TSEOptions.aspx?ReportID=2&Referrer=TSEReports.aspx&DataSource=NTD (accessed March 1, 2014).

Jaffe, A. May 7, 2013. "Testimony of Amy Myers Jaffe to the House Committee on Energy and Commerce." *UCDAVIS*, http://gsm.ucdavis.edu/sites/main/files/file-attachments/househearingsjaffeusenergyexportsandforeignpolicyupdate.pdf

Kenny, C. January 23, 2014. "Factory Jobs Are Gone. Get Over It." *Bloomberg Businessweek*.

Levinson, M. April 15, 2013. "Hollowing Out of U.S. Manufacturing: Analysis and Issues for Congress." *Congressional Research Service,* http://www.fas.org/sgp/crs/misc/R41712.pdf

Mandel, M. January–February 2012. "The Myth of American Productivity." *The Washington Monthly*, https://www.google.com/webhp?source=search_app#q=the+myth+of+american+productivity

Mandel, M., and D.G. Carew. November 2011. "How a Competitive Audit Can Help Create Jobs." *Progressive Policy Institute*, http://progressivepolicy.org/wp-content/uploads/2011/11/11.2011-Mandel-Carew_How-A-Competitiveness-Audit-Can-Help-Create-Jobs.pdf

McCormack, R. December 29, 2009. "The Plight of American Manufacturing." *The American Prospect*, http://prospect.org/article/plight-american-manufacturing

McCormack, R.A. January 16, 2014. "Top Government Officials Discover There Is Little Left Of America's Telecom Equipment Manufacturing Industry; United States Has To Create A New Industry From Scratch." *Manufacturing and Technology News*, http://www.manufacturingnews.com/news/Defense-Production-Act-Committee-telecom-0116141.html

Mellman Group and North Star Opinion Research. February 3, 2014. "New National Poll: Voters Blame Lawmakers for Weak Job Growth." *American Manufacturing*, http://americanmanufacturing.org/blog/new-national-poll-voters-blame-lawmakers-weak-job-growth

Miroff, N. July 1, 2013. "With Mexican Auto Manufacturing Boom, New Worries." *Washington Post*, http://www.washingtonpost.com/world/the_americas/with-mexican-auto-manufacturing-boom-new-worries/2013/07/01/10dd57e8-d7d9-11e2-b418-9dfa095e125d_story.html

MIT. 2013. "Report of the MIT Taskforce on Innovation and Production." *MIT News*, http://web.mit.edu/press/images/documents/pie-report.pdf

Morici, P. 1997. *Barring Entry? China and the WTO*, http://www.rhsmith.umd.edu/faculty/pmorici/cv_pmorici.htm

Morris, C.R. 2013. *Comeback: America's New Economic Boom*. New York, NY: Public Affairs.

Morrison, W.M., and M. Labonte. January 9, 2008. "China's Currency: Economic Issues and Options for U.S. Trade Policy." *Congressional Research Service*. http://www.fas.org/sgp/crs/row/RL32165.pdf

Morrison, W.M., and M. Labonte. July 22, 2013. "China's Currency Policy: An Analysis of the Economic Issues." *Congressional Research Service*, https://www.fas.org/sgp/crs/row/RS21625.pdf

Morse, E.L. March 20, 2012. *Energy 2020: North America, the New Middle East?*, http://www.morganstanleyfa.com/public/projectfiles/ce1d2d99-c133-4343-8ad0-43aa1da63cc2.pdf

NBER. August 20, 2010. *U.S. Business Cycle Expansions and Contractions*, http://www.nber.org/cycles/cyclesmain.html

New York Times. August 14, 1985. "Berkshire to Shut Textile Division." *New York Times*, http://www.nytimes.com/1985/08/14/business/berkshire-to-shut-textile-division.html

The New York Times. November 9, 1993. "The 'Great Debate' Over NAFTA." http://www.nytimes.com/1993/11/09/opinion/the-great-debate-over-nafta.html

New York Times. November 12, 1993. "Business Lobbying for Trade Pact Appears to Sway Few in Congress." http://www.nytimes.com/1993/11/12/us/business-lobbying-for-trade-pact-appears-to-sway-few-in-congress.html

Nordhaus, W. May 2005. "The Sources of the Productivity Rebound and the Manufacturing Employment Puzzle." *National Bureau of Economic Research*, http://www.nber.org/papers/w11354

Norris, F. January 15, 2011. "Manufacturing Is Surprising Bright Spot in U.S. Economy." *New York Times,* http://www.nytimes.com/2012/01/06/business/us-manufacturing-is-a-bright-spot-for-the-economy.html?_r=0

OANDA. 2014. *Historical Exchange Rates,* http://www.oanda.com/currency/historical-rates/

Office of the Deputy Assistant Secretary of Defense. October 2013. *Annual Industrial Capabilities Report to Congress,* http://www.acq.osd.mil/mibp/docs/annual_ind_cap_rpt_to_congress-2013.pdf

Pew Research Center. 2014. "Three Technology Revolutions." *PewResearch Internet Project,* http://www.pewinternet.org/three-technology-revolutions/

Pierce, J.R., and P.K. Schott. April 2014. "The Surprisingly Swift Decline of U.S. Manufacturing Employment." *Federal Reserve,* http://www.federalreserve.gov/pubs/feds/2014/201404/201404pap.pdf

Pro Mexico. March 3, 2014. *Delphi Brings Mexico to Megatrends,* http://negocios.promexico.gob.mx/english/04-2013/art03.html

Reich, R.B. May 28, 2009. "Manufacturing Jobs Are Never Coming Back." *Forbes,* http://www.forbes.com/2009/05/28/robert-reich-manufacturing-business-economy.html

Reuters. June 27, 2013. "GM Is Spending A Lot of Money to Expand Its Production in Mexico." *Business Insider,* http://www.businessinsider.com/gm-invests-691-million-in-mexico-plants-2013-6

Reshoring Initiative. 2014. "TCO Estimator and Case Study Account Creation." http://www.reshorenow.org/signup.cfm

Richman, S.L. May 30, 1988. "The Reagan Record on Trade: Rhetoric vs. Reality." *The Cato Institue,* http://www.cato.org/publications/policy-analysis/reagan-record-trade-rhetoric-vs-reality

Riffkin, R. February 17, 2014. "Unemployment Rises to Top Problem in the U.S." *Gallup,* http://www.gallup.com/poll/167450/unemployment-rises-top-problem.aspx

Rothstein, J., and R.E. Scott. October 1997. *The Cost of Trade With China – Women and Low-Wage Workers Hit Hardest by Job Losses in All 50 States,* http://www.epi.org/publication/issuebriefs_ib122/

Rothstein, J., and R.E. Scott. September 19, 1997. "NAFTA and the States—Job Destruction is Widespread." *Economic Policy Institute,* http://www.epi.org/publication/issuebriefs_ib119/

Saez, E. September 3, 2013. *Striking It Richer: The Evolution of Top Incomes in the United States,* http://elsa.berkeley.edu/~saez/saez-UStopincomes-2012.pdf

Scott, R.E. March 1, 1996. "North American Trade After NAFTA." *Economic Policy Institute* http://www.epi.org/publication/epi_virlib_briefingpapers_1996_northa/

Scott, R.E. November 1, 1999. NAFTA's Pain Deepens—Job Destruction Accelerates in 1999." *Economic Policy Institute,* http://www.epi.org/publication/briefingpapers_nafta99_nafta99/

Scott, R.E. February 1, 2000. "The High Cost of the China-WTO Deal—Administration's Own Analysis Suggests Spiraling Deficits, Job Losses." *Economic Policy Institute,* http://www.epi.org/publication/issuebriefs_ib137/

Scott, R.E. October 9, 2007. "Costly Trade with China—Millions of U.S. Jobs Displaced With Net Job Loss in Every State." *EPI Briefing Paper,* http://s1.epi.org/files/page/-/old/briefingpapers/188/bp188.pdf

Scott, R.E. March 23, 2010. "Unfair China Trade Costs Jobs." *Economic Policy Institute,* http://s4.epi.org/files/page/-/bp260/bp260.pdf

Scott, R.E. February 26, 2014. "Stop Currency Manipulation and Create Millions of Jobs ." *Economic Policy Institute,* http://www.epi.org/publication/stop-currency-manipulation-and-create-millions-of-jobs/

Shaiken, H. March 3, 2010. "Commitment is a Two Way Street: Toyota, California and NUMMI." http://www.treasurer.ca.gov/nummi/report.pdf

Sirkin, H.L., M. Zinser, and D. Hohner. August 2011. *Made in America, Again.* Boston, MA: Boston Consulting Group, http://www.bcg.com/documents/file84471.pdf

Sirkin, H.L. September 30, 2011. "Is U.S. Manufacturing Making a Comeback?" *HBR Blog Network,* http://blogs.hbr.org/2011/09/us-manufacturing-comeback/

Tonelson, A. April 9, 2013. "An Industrial Has-Been? Chinese Imports Keep Winnng Market Share in High-Value Industry." *U.S. Business And Industry Council,* http://www.americaneconomicalert.org/Chinaimportpenetrationreport30132.pdf

Tonelson, A. January 2013. "Import Penetration Rises Again in 2011; Challenges Manufacturing Renaissance, Insourcing Claims." *U.S. Business and Industry Council,* http://americaneconomicalert.org/USBICImportPenetrationReport2013Final.pdf

U.S. Census Bureau. 2013. *Manufacturers' Shipments, Inventories, and Orders,* http://www.census.gov/manufacturing/m3/historical_data/index.html.

U.S. Congress. *Congressional Record.* 106th Cong., 2nd sess., 2000. Vol. 146, pt. 7. http://books.google.com/books?id=z6E993ZslkAC&pg=PA9129&lpg=PA9129&dq=china+pntr+congressional+record+may+2000&source=bl&ots=qdd8Oln5ht&sig=UH2_wkCTb4WdQIR86B_JUVCQm0E&hl=en&sa=X&ei=bQEfU_P7OoO3kAfr_YBg&ved=0CCcQ6AEwAA#v=onepage&q=china%20pntr%20congressi

U.S. Department of Agricultural. 2002. "Charts and Maps: U.S. Number of Farms and Workers." *National Agricultural Statistics Service,* http://www.nass.usda.gov/Charts_and_Maps/Farm_Labor/fl_frmwk.asp; http://www.nass.usda.gov/index.asp

U.S. Department of Agricultural. September, 2013. "Agricultural productivity in U.S." *Economic Research Service,* http://www.ers.usda.gov/data-products/agricultural-productivity-in-the-us.aspx#.UsbGANJDvmc.

U.S. Food and Drug Administration. June 20, 2011. *Pathway to Global Product Safety and Quality,* http://www.fda.gov/downloads/aboutfda/centersoffices/oc/globalproductpathway/ucm259845.pdf

U.S. International Trade Commission. September 1999. *Assessment of the Economic Effects on the United States of China's Accession to the WTO,* http://www.usitc.gov/publications/docs/pubs/332/PUB3229.PDF

U.S. Senate Committee on Armed Services. November 8, 2011. *The Committee's Investigation into Counterfeit Electronic Parts in the Department of Defense Supply Chain,* http://www.gpo.gov/fdsys/pkg/CHRG-112shrg72702/pdf/CHRG-112shrg72702.pdf

USA Today. November 13, 2013. *Big Milestone: U.S. Producing More Oil Than it Imports* http://www.usatoday.com/story/news/nation/2013/11/13/us-oil-production-exceeds-imports/3518245/

USDA. 2012. *Economic Research Service Using Data from USDA, National Agricultural Statistics Service, Census of Agricultural and Farms, Land in farms, and Livestock Operations: 2012 Summary.* USDA.

Ventola, C.L. November 2011. "The Drug Shortage Crisis in the United States." *National Center for Biotechnology Information,* http://www.ncbi.nlm.nih.gov/pmc/articles/PMC3278171/

Webber, M. 2009. "Erosion of the Defense Industrial Support Base." In *Manufacturing a Better Future For America,* ed. R. McCormack, 245–280. Washington, D.C.: The Alliance for American Manufacturing.

Worstall, T. February 3, 2012. "That Giant Sucking Sound of Manufacturing Jobs Going to China." *Forbes,* http://www.forbes.com/sites/timworstall/2012/02/03/that-giant-sucking-sound-of-manufacturing-jobs-going-to-china/

World Trade Organization. 2014. "Dispute Settlement: The Disputes." http://www.wto.org/english/tratop_e/dispu_e/find_dispu_cases_e.htm

Yudken, J.S. September 2010. *Manufacturing Insecurity – America's Manufacturing Crisis and the Erosion of the U.S. Defense Industrial Base,* http://www.highroadstrategies.com/downloads/HRS-IUC-Def-Industrial-Base-Report.pdf

Index

OTHER TITLES FROM THE ECONOMICS COLLECTION

Philip Romero, The University of Oregon and Jeffrey Edwards,
North Carolina A&T State University, Editors

- *Managerial Economics: Concepts and Principles* by Donald Stengel
- *Your Macroeconomic Edge: Investing Strategies for the Post-Recession World* by Philip J. Romero
- *Working with Economic Indicators: Interpretation and Sources* by Donald Stengel
- *Innovative Pricing Strategies to Increase Profits* by Daniel Marburger
- *Regression for Economics* by Shahdad Naghshpour
- *Statistics for Economics* by Shahdad Naghshpour
- *How Strong Is Your Firm's Competitive Advantage?* by Daniel Marburger
- *A Primer on Microeconomics* by Thomas Beveridge
- *Game Theory: Anticipating Reactions for Winning Actions* by Mark L. Burkey
- *A Primer on Macroeconomics* by Thomas Beveridge
- *Economic Decision Making Using Cost Data: A Guide for Managers* by Daniel Marburger
- *The Fundamentals of Money and Financial Systems* by Shahdad Naghshpour
- *International Economics: Understanding the Forces of Globalization for Managers* by Paul Torelli
- *The Economics of Crime* by Zagros Madjd-Sadjadi
- *Money and Banking: An Intermediate Market-Based Approach* by William D. Gerdes
- *Basel III Liquidity Regulation and Its Implications* by Mark A. Petersen and Janine Mukuddem-Petersen

Announcing the Business Expert Press Digital Library

*Concise E-books Business Students Need
for Classroom and Research*

This book can also be purchased in an e-book collection by your library as
- a one-time purchase,
- that is owned forever,
- allows for simultaneous readers,
- has no restrictions on printing, and
- can be downloaded as PDFs from within the library community.

Our digital library collections are a great solution to beat the rising cost of textbooks. E-books can be loaded into their course management systems or onto students' e-book readers.

The **Business Expert Press** digital libraries are very affordable, with no obligation to buy in future years. For more information, please visit **www.businessexpertpress.com/librarians**. To set up a trial in the United States, please email **sales@businessexpertpress.com**.

www.ingramcontent.com/pod-product-compliance
Lightning Source LLC
Chambersburg PA
CBHW060330200326
41519CB00011BA/1888